This is a study and translation of the *Zhou bi suan jing,* a Chinese work on astronomy and mathematics which reached its final form around the first century AD. The author provides the first easily accessible introduction to the developing mathematical and observational practices of ancient Chinese astronomers and shows how the generation and validation of knowledge about the heavens in Han dynasty China related closely to developments in statecraft and politics.

NEEDHAM RESEARCH INSTITUTE STUDIES · I

Astronomy and mathematics in ancient China: the *Zhou bi suan jing*

The Needham Research Institute Studies series will publish important and original new work on East Asian culture and science which develops or links in with the publication of the *Science and Civilisation in China* series. The series will be under the editorial control of the Publications Board of the Needham Research Institute.

Astronomy and mathematics in ancient China: the *Zhou bi suan jing*

周髀算經

Christopher Cullen

Senior lecturer in the History of Chinese Science and Medicine, School of Oriental and African Studies, University of London

Published by the Press Syndicate of the University of Cambridge
The Pitt Building, Trumpington Street, Cambridge CB2 1RP
40 West 20th Street, New York, NY 10011–4211, USA
10 Stamford Road, Oakleigh, Melbourne 3166, Australia

First published 1996

Printed in Great Britain
at the University Press, Cambridge

A catalogue record for this book is available from the British Library

Library of Congress cataloguing in publication data

Cullen, Christopher.
Astronomy and mathematics in ancient China: the Zhou bi suan jing/
Christopher Cullen.
p. cm.
Includes bibliographical references and index.
ISBN 0 521 55089 0 (hc)
1. Mathematics, Chinese. 2. Astronomy, Chinese. I. Chou pi suan ching. English.
II. Title.
QA27.C5C85 1996
510′.931–dc20 95-32979 CIP

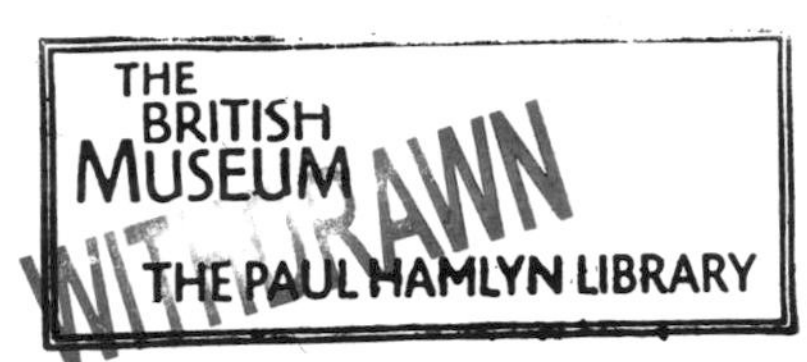

Contents

This book is dedicated with respect and gratitude
to D. C. Lau, for holding up more than one corner,
and to Ho Peng-Yoke, for bringing fuel in snowy weather.

Preface

The *Zhou bi* 周髀 is a collection of ancient Chinese texts on astronomy and mathematics. It was probably assembled under the Western Han 漢 dynasty during the first century BC, although it was traditionally reputed to have been written in Western Zhou 周 times about a thousand years earlier. Following the explanation given in the book itself, the title may be rendered 'The Gnomon of the Zhou [dynasty]'.[1] Much of the text consists of calculations of the dimensions of the cosmos using observations of the shadow cast by a simple vertical pole gnomon. The *Zhou bi* follows the doctrine of the *gai tian* 蓋天 cosmography, in which an umbrella-like heaven rotates about a vertical axis above an essentially plane earth.[2] It is therefore unique in being the only rationally based and fully mathematicised account of a flat earth cosmos.

In its present form the *Zhou bi* certainly deserves the title of 'the principal surviving document of early Chinese science' given to it by A. C. Graham in his *Later Mohist Logic, Ethics and Science*. As to its value, opinions have differed. A twentieth–century Japanese scholar well trained in modern science has called it 'a veritable golden treasury of knowledge for all ages'.[3] On the other hand an early seventeenth–century

[1] See paragraph #B15. All references to my translation take this form, in which the letter designates the sections into which I have divided the text, and the figure gives the number of the paragraph within the section. In the main translation each paragraph number is followed by a note of where the original text begins in the edition of Qian (1963). Thus 34d refers to page 34, the fourth column counted from the right. *Zhou bi* is the original title of the book, but after the work became a university set text in the seventh century AD the words suan jing 算經 'mathematical classic' were usually added. In Wade-Giles romanisation the title will be found as *Chou pi suan ching*, or sometimes as *Chou pei suan jing*. The reading of the second character as pinyin *bi,* Wade-Giles *pi* is supported by the earliest authority, who is Li Ji 李籍 writing in the eleventh century AD: see the details given in chapter 2.

[2] My use of the term 'cosmography' rather than 'cosmology' is a deliberate distinction. By the first of these terms I mean a description that (at least ostensibly) is mainly concerned with the shape and size of the heavens and the earth, and with the disposition and motions of the heavenly bodies – a cosmic equivalent of geography. By the second term I mean any theory of how the universe works in a more metaphysical sense. In China I would call discussion of Yinyang and Five Phase thinking cosmology in this sense. Of course both terms involve prefabricated 'observer categories', and we cannot guarantee that they will correspond to the 'actor categories' we hope to recover by studying the writings of ancient Chinese thinkers. In a case such as Plato's *Timaeus* the cosmographical/cosmological distinction hardly seems to be present in the author's mind at all. But in my experience this initial distinction is a helpful one, even if it may need revision at a later stage.

[3] Nōda (1933), 140.

Chinese scholar, also well acquainted with the Western science of his own day, dismissed much of the book as 'the greatest stupidity of all time'.[4] The aim of this study is to make the *Zhou bi* accessible to anybody with an interest in the history of science, or in the history of Chinese culture, so that they can judge for themselves.

There is much to be said for reading a translation with a fresh mind, without allowing one's first impression to be influenced by the introductory discussion in which (with perhaps too much proprietorial zeal) the translator tells the reader what the ancient writer really meant, and why he wrote his book. While the language of the *Zhou bi* is usually straightforward enough, I think that most readers will find that an inspection of the plain text will give them no more than a general and rough impression of the concerns of the work. To supply the sense of context and purpose needed to understand the structure and content of the work, I have therefore provided substantial discussions of relevant issues.

First of all, I have given a brief survey of as much of the history of early Chinese astronomy as is necessary to understand the *Zhou bi*. There is at present no introductory study in a Western language suitable for this purpose.[5] With this essential foundation laid down I have then discussed the main features of the text of the *Zhou bi*, and analysed its principal themes. Next I have looked into the question of when and how the work originated, and discussed its later history, including questions of textual transmission. After the translation itself there follow, in conclusion, translations of four important short essays written by the first commentator on the *Zhou bi* around the late third century AD.

I make no claim to have deployed any very sophisticated methodology in this book. Half the effort required was of the old–fashioned philological and text-critical variety. Otherwise I had three main guiding principles before me, which I think would be common to all serious historians of science nowadays. I have tried firstly to reconstruct the thought and practice of the ancient authors from their writings rather than taking it for granted that modern categories will apply. This is of course an aim impossible of complete attainment, as my use of such terms as 'astronomy' and 'mathematics' has already demonstrated. Secondly I have taken it as my first assumption that the various things people did in the past made sense to them as part of a more or less systematised approach to the tasks they were addressing. Naturally, in the end they may turn out to have had the usual human capacity for confusion or self-delusion. Lastly, I have tried

[4] Xu Guangqi 徐光啟 (baptised as Paul Xu), quoted in Dai Zhen's 戴震 introduction to the edition of the *Zhou bi* in the *Siku quanshu* collection: photographic reprint (1983) vol. 786, 2.

[5] The principal problem with the wide-ranging and scholarly treatment of astronomy in Needham (1959) is that it deliberately discounts the calendrical astronomy which is such an important element in the *Zhou bi*. For a general critique, see Cullen (1980).

as far as possible not to draw boundaries which may not have seemed at all solid at the time, or to force issues which had not yet occurred to anybody. How far I have succeeded must be left for others to judge.

I do not need to find arguments to persuade the student of the history of Chinese science to pay attention to the *Zhou bi*, nor will such scholars need to be convinced of the need for an adequate treatment of this work in a Western language.[6] Graham's estimation of the importance of this text is clearly correct, and in addition the *Zhou bi* has the added interest of representing the work of an unofficial group of astronomical thinkers, whose ideas may give us an alternative to the otherwise overwhelmingly official bias of Chinese astronomical literature.

But what can a historian of Western science hope to gain from the encounter? At the most trivial level, the *Zhou bi* is certainly a novelty for someone from an intellectual world where the only flat-earthers are Pre-Socratics known through a few doxographic fragments, or dogged eccentrics such as Cosmas Indicopleustes. More seriously, ancient Chinese astronomical theory and practice appears to have had no significant connections with the worlds of Hellas, Babylon or Egypt. For once, therefore, we can cite parallels and differences without being drawn into the old conflicts about who influenced whom when and to what extent. The astronomy of Han China is an independent test case. If we make proper use of the evidence it gives us we may be able to make a little progress towards deciding how far our perceptions are dictated by an objective natural world 'out there', and how far our world-view is a cultural construct. Not the smallest challenge presented by the *Zhou bi* is that we have to think away the concept of the celestial sphere that has been natural to all scientifically educated Westerners since at least the time of Eudoxus – a transition much more difficult to make than the switch from a spherical to a flat earth. Such exercise is good for the mind, and sharpens historical perception.

At some point in a preface an author has to pay his intellectual debts: as Isaac D'Israeli says, 'gratitude is not a silent virtue'. I am however in a slightly odd position in this respect. Through the peculiarities of my initially amateur interests in these matters, and through the imaginative tolerance of the Governing Body of the School of Oriental and African Studies (SOAS), I first read the *Zhou bi* at a time when I knew only classical Chinese. Modern Chinese and Japanese scholarship were initially closed to me. I was therefore in the fortunate position of forming at least a general view of the text fairly independently. Later, of course, I was able to read the work of Nōda Churyō 能田忠亮 and Qian Baocong 錢寶琮 on the *Zhou bi*, and I realised how much

[6] The pioneering French translation of Biot (1841) was successful in revealing many of the main features of the text, but has now been left behind by developments in scholarship. The treatment of the cosmography of the *Zhou bi* in Chatley (1938) is heavily dependent on Biot and comes to some very misleading conclusions. It is unfortunately followed by Needham (1959), 212. Chatley's views are spared criticism in the otherwise excellent short review in Nakayama (1969), 25–35.

more there was to many problems than I had thought. I do however feel some confidence in the basic views formed in my initial state of relative ignorance, since I found most of them confirmed in the work of these two great East Asian scholars. My debt to them is so large that it would be tedious to footnote every instance where I have been influenced by their work.

Over the years I have discussed the *Zhou bi* with many friends and colleagues. First place must go to my PhD supervisor at SOAS, D. C. Lau, who introduced me to the text and (I now realise) worked harder to teach me the rudiments of classical scholarship than I had any right to expect. In the next office to D. C. Lau was Angus Graham, who also gave generously of his time. During autumn 1991 I enjoyed the hospitality of the Institute of History at National Tsing-Hua University, Taiwan while completing the first draft of this book, and had the benefit of long conversations with Professor Huang Yi-Long and Professor Fu Daiwie. A draft of the text was read and criticised by Professor Nathan Sivin and Professor Geoffrey Lloyd, and I am most grateful to both of them for giving their time to this task. The responsibility for all remaining mistakes and omissions is of course my own. Finally, of course, only my wife Anne Farrer and my sons Peter and Robert are qualified to say how much my authorial efforts have cost those around me. To them goes my deep gratitude for their continued tolerance and support.

1

The background of the *Zhou bi*

I shall argue that the *Zhou bi* can best be understood as a product of the Han 漢 age. The Han dynasty lasted from 206 BC to AD 220, with a relatively brief interruption from AD 9 to 23, when a powerful courtier, Wang Mang 王莽, usurped the throne. He was the only ruler of his short-lived Xin 新 'New' dynasty. It is usual to call the first half of the dynasty 'Western Han' and the second half 'Eastern Han', since the capital shifted from Chang'an 長安 in the west to Loyang 洛陽 in the east after the Xin interregnum. The Han dynasty therefore spanned an interval as long as that which separates late twentieth century England from the time of Queen Elizabeth the First.

Given such a long period of time, it is clear that we must be cautious in generalising about Han culture. Of course the analogy with English history is a little misleading. The Chinese world of the second century BC was not separated from the world of the second century AD by anything comparable with the scientific and industrial revolutions, the rise and dissolution of the British Empire, and the world-wide military and social cataclysms of the twentieth century. Nevertheless the Han writers whose words have come down to us are clearly conscious of living in a world where social, political and intellectual change are continual, sometimes for the better and sometimes for the worse.

The most striking evidence of change for a Han historian was the contrast between his own age and that which had preceded it. The world in which he lived was a political unity bound together by ideological and administrative structures unprecedented in China. There had been previous kingdoms claiming rule over the whole country, such as the Shang 商 in the five centuries up to about 1025 BC, or the earlier centuries of the rule of the Zhou 周 who overthrew them. But after their capital fell to outside enemies in 771 BC the Zhou kings were no more than nominal rulers. The real power was in the hands of their vassals, the lords of the great feudal states, some comparable in size with modern European countries. By the time of Confucius around 500 BC warfare between the states was frequent, and it rose in a crescendo of scale and intensity until the western state of Qin 秦 destroyed all its rivals and set up a unified empire in 221 BC.

Rejecting the old Zhou system of feudal appanages, the first ruler of Qin extended the civil administrative system of his own state to the whole country, so that for the first time China was ruled by a non-hereditary bureaucracy. He was guided in this by a varied group of statesmen, who generally advocated a ruthless *realpolitik* in which the preservation of the ruler's power and the promotion of his interests came before all else. Chinese historians of later centuries, perhaps over-concerned to systematise the past, tended to see these men as representatives of a self-conscious ideological movement, which they labelled retrospectively as the *Fa jia* 法家 'School of Laws' or 'Legalists'. The first ruler of Qin saw his conquest in cosmic terms. The world was governed by the cyclical dominance of the Five Powers *wu de* 五德 of Earth, Wood, Metal, Fire, Water (later commonly referred to as the *wu xing* 五行 'five phases') and he believed that it was through the rise of Water that he had come to power. Lucid discussions of five phase theory will be found in Graham (1989) and Sivin (1987), and the details need not detain us here since the theory does not appear in the *Zhou bi*..

But the harsh rule of Qin was short-lived. The death of the first Qin emperor in 210 BC was followed by widespread rebellion, from which by 202 BC one of the many contenders emerged with sufficient clear advantage to proclaim himself emperor of a new dynasty, the Han. This was Liu Bang 劉邦, originally a petty official of peasant origins who is better known to history by his posthumous title Gao Zu 高祖 'The High Ancestor'.

The Han dynasty mostly continued the political and administrative structures of the Qin. In the field of ideas, however, the situation was different. The age of the so-called 'Warring States' from the fifth century BC to the rise of Qin had been the most varied and fruitful period of Chinese thought, and, unlike the Qin, the Han had the time and security to create a synthesis to serve its needs. The intellectual products of an age of semi-anarchy were transmuted into an ideology to underpin the new imperial state. By the first century BC the main lines of a new world-view had been laid down on the ostensible basis of the ancient texts preserved by the scholars known as *Ru* 儒. This is the group usually referred to by Westerners as 'Confucians', although it is doubtful how far Confucius would have recognised his own voice in the doctrines taught by the professors appointed to supervise the intellectual formation of imperial bureaucrats.

He would certainly not have recognised the elaborate cosmology which was created to underpin the moral and political orthodoxy of the centralised empire. This was the age when the ideas of Yin 陰 and Yang 陽 and the Five Phases, and of the Book of Change, *Yi Jing* 易經, were woven into complex schemes in which every phenomenon found a place in an ordered cosmos, at the centre of which the Chinese emperor had the duty of holding the proper balance between Heaven, Earth and Man. Since the appearance of the sky was one of the clearest and most striking expressions of cosmic order (or the lack of it), it is not surprising that an important department of the imperial bureaucracy was concerned with astronomy. It is within the context of the marked changes in astronomical theory and practice under the Han that we must try to understand the nature and origins of the *Zhou bi.*

What follows is an attempt to give the reader without specialist knowledge an outline understanding of the place of astronomy in the culture of the early imperial period. It is not intended as a complete introduction to early Chinese astronomy, and for the sake of brevity some points are therefore made in summary form.

More detailed information on the development of Chinese astronomy from early times up to the seventeenth century will be found in Needham (1959). Despite the criticisms of Cullen (1980) this remains the best general account of the topic in a European language, and its references to earlier studies are comprehensive. For the present purpose Needham's low estimate of the importance of calendrical astronomy in China is however a significant defect. Sivin (1969) is therefore an essential supplement for anyone who wants to get a detailed picture of how the users of the early astronomical systems actually did their calculations. Yabuuchi (1969) collects (although it does not synthesise) the researches of the twentieth century's greatest East Asian scholar in the field of Chinese astronomy. There are a number of general histories of Chinese astronomy in Chinese. Anon. (1987) is a convenient one-volume treatment whose authors are more scrupulous than is usual in giving references to original sources and the research literature.

The foundation charter

The Book of Genesis stands at the beginning of the sacred canon of the Judaeo-Christian tradition. In its first chapter we are told of God's creation of the natural order, and in particular of the creation of man. By contrast, in the earliest canonical account of the past in the Confucian tradition we are told with equal solemnity of the creation of the socio-political order by a human ruler. According to the first chapter of the *Shu jing* 書經 'Book of Documents', perhaps compiled around 400 BC, an important part of this process was the commission given to two hereditary lineages of star-clerks by the legendary Emperor Yao 堯 in remote antiquity:

> Thereupon he ordered Xi 羲 and He 和 to accord reverently with august Heaven, and its successive phenomena, with the sun, the moon and the stellar markers, and thus respectfully to bestow the seasons on the people.[7]

Xi and He each have two sons, who are dispatched to the four quarters of the earth with separate commissions to note the times of the summer and winter solstices, and of the spring and autumn equinoxes. Finally the Emperor turns back to the fathers:

[7] For text and a slightly different translation see Karlgren (1950), pp. 2 and 3.

> O you Xi and He! The period is of three hundreds of days, and six tens of days, and six days. Use intercalary months to fix the four seasons correctly, and to complete the year.[8]

This document has sometimes been called the foundation charter of Chinese astronomy. Whatever its historical value, it certainly does give an early indication of the importance attached to calendrical astronomy by the pre-modern Chinese state. Until the end of imperial China all governments maintained a staff of astronomical specialists. While such officials were not at the top of the bureaucratic tree, no government felt it could manage without them. We must now say something about the functions that belonged to their office.

The role of astronomy

Calendrical astronomy

The essential task given to Xi and He is 'respectfully to bestow the seasons on the people'. They are to prepare a calendar, which will be promulgated with imperial authority. This task was a principal concern of the official astronomical establishment for all the centuries during which it is known to have existed. The basic need seems obvious enough. Clearly no government of any complexity can operate without some means of unambiguously designating in advance the days on which it expects to perform certain functions, or requires certain obligations to be fulfilled. But to the reader familiar only with the Gregorian system now used in most countries, the preparation of a calendar may seem a puzzlingly trivial task to absorb the energies of an important group of bureaucrats. In fact the problems encountered could be highly complex, and might demand ingenuity of a high order for their solution.

A modern Gregorian style calendar gives the days of the twelve months in numbered sequence, and indicates on which days of the seven day week they fall. Once the rules for the placing of leap years are understood, a calendar for any given year can easily be drawn up. The information required is simple: one must know the names of the months in order, and the numbers of days in each (allowing for the extra day for February in a leap year), and the names and order of the days of the week. Knowledge of the day of the week for any given day of a given month in a given year enables the two sequences to be matched up for all other dates.

As we shall see, the luni-solar nature of the ancient Chinese calendar makes the equivalent task much more difficult. The Chinese astronomers had to work with real months that followed lunations quite closely, as well as keeping a civil year of a whole number of months in step with the seasons. But in addition, from near the beginning

[8] Same reference.

of the imperial age astronomers had other tasks imposed on them. It was expected that the motions of the visible planets should be tabulated in detail, and that lunar eclipses should be predicted. More difficult still, some advance warning of the possibility of solar eclipses was expected. To do all this, it was essential to have a complete system of mathematical astronomy in operation. Some of the details of such a system are discussed in the *Zhou bi*. But why was all this effort considered worthwhile?

The timekeeping function

It is likely that some Chinese astronomical specialists were mainly motivated by intellectual curiosity, or perhaps by a wish to rise in their careers as successful professionals. Such reasons do not explain why such specialists were felt to be an essential part of the state apparatus. One explanation frequently offered by both Chinese and Western authors runs along the following lines. China was an agrarian civilisation. Successful agriculture demands that such operations as sowing and harvesting are carried out at the proper time. Therefore it was essential for the Chinese government to provide the peasantry with an accurate calendar, and indeed to make constant efforts to improve its accuracy.

This explanation seems to me both misleading and inadequate. In the first place it is not much of a compliment to the Chinese peasant to suggest that without the help of the government he would not have known when conditions were appropriate for the work he had to do. Further, given the natural vagaries of the weather and variations in local conditions, it is likely that a peasant who sowed and reaped by the calendar would do rather worse than one who trusted to observation and experience. Then we have to face the fact that only a small part of the work of official astronomers related directly to the passage of the seasons at all: once the dates of solstices have been determined there is in principle nothing left to say on the topic. And even then, for agricultural purposes a few days' error about (say) the date of midsummer is neither here nor there. The elaborate attention paid to the movements of the moon and the planets is, according to the view criticised here, quite inexplicable.

The ritual function

Why then should Chinese astronomers have wanted precise answers to such questions as when a lunation began? It will help if we recall the Chinese emperor's role as the pivotal element linking the human microcosm and the natural macrocosm. He is responsible for the orderly functioning of both spheres. Visible disorder in nature is a sign of a malfunction in the human order, and may thus evoke direct criticism of the emperor's rule. In the natural order celestial portents drew the most attention, and it is therefore quite understandable that Chinese astronomers tried to reduce to rule as many astronomical phenomena as possible, with the ultimate aim of predicting everything predictable. If, for instance, the occurrence of a lunar eclipse could be predicted its significance as a portent was much reduced. More positively, the image of the emperor as successful preserver of the cosmic order was inevitably enhanced if his government was seen to comprehend the subtlest motions of the heavens.

Such motivations go some way towards explaining the astronomical preoccupations of the Chinese state. But once the astronomers had produced their detailed tabulations, did they answer any purpose beyond serving as imperial status symbols? Why should anyone care that the moment of winter solstice fell precisely when it did and not an hour or so later? In fact such data were seen as of the highest practical importance, both for the state and for the individual.

For the state, it was necessary that certain imperial rituals should be carried out at the proper times. Such rituals, often involving the emperor himself as celebrant were an essential contribution to the maintenance of cosmic order. If they were mistimed, they could fail to produce benefit or even do harm. If winter solstice fell a few minutes before midnight, but the astronomers predicted it an hour later, the emperor would be led to carry out his sacrificial ritual a whole day late. For the individual the consequences were similar. Almost all the activities of daily life, from taking a wife to closing a business deal, were conducted according to an elaborate divinatory scheme of lucky and unlucky days. Such schemes were an essential part of the calendar in the form most widely distributed. Commercially published almanacs giving this information are still bought in huge numbers by modern Chinese people. If the calendar is in error, bad fortune or even serious danger might be the result.

The role of portents

It appears, then, that a great deal of what we would nowadays characterise as scientific activity was supported by the Chinese state for motives that had little to do with modern science. For completeness I will mention a related aspect of ancient Chinese astronomical activity. This was the observation of transient phenomena such as comets, novae and meteor showers. None could be predicted, and all were therefore potentially ominous. As a result it fell to the official astronomers to take careful note of all such phenomena, and of the precise times and locations at which they occurred, with the aim of ensuring that their significance was correctly deduced and reported to the throne. Many such records have been preserved to the present day, and remain of the greatest value to modern astronomers. The *Zhou bi* does not however concern itself with this aspect of official astronomical activity. In this it follows a division between portent astrology *tian wen* 天文'celestial patterns' and mathematical astronomy *li fa* 曆法 'calendrical methods' which is basic to ancient literature dealing with the heavens.

The problem of the calendar

Modern astronomical theory

Quite apart from the fact that the Chinese calendrical astronomer was expected to produce detailed predictions of a wide range of phenomena, it has already been mentioned that the basic nature of the Chinese calendar set him a more difficult task than running the Gregorian calendar demands. This was because the Chinese calendar was of the luni-solar type, which uses the moon as well as the sun as an important time-marker. In Europe the moon plays no role in the civil calendar at all, although it is still important in fixing the date of Easter in the calendar of the Christian church. The following sections introduce the problems of running a luni-solar calendar, so far as they are relevant to the concerns of the *Zhou bi*. I begin with a very short reminder of some basic astronomical facts, which may be skipped by those who do not need reminding. Fuller and very clear introductions to basic naked-eye astronomy, more or less from scratch, will be found in Kuhn (1957) and Dicks (1970).

On the modern view, the earth is an almost exact spheroid of radius 6400 km which rotates once daily on its axis. Once a year it completes a revolution round the sun. Its axis of daily rotation is tilted 23.5 degrees from the perpendicular to the plane of its orbit round the sun. During the course of a year the earth's tilted axis points in an almost unchanged direction in space, stabilised by the gyroscopic action of the spinning earth. Due to the effect of the sun's gravitational pull on the earth's slight equatorial bulge, over a longer period it becomes clear that this axis is in fact precessing conically like the axis of a child's top, so that it takes about 23 000 years to return to its original orientation.

As the earth moves in its orbit round the sun, the moon orbits round the earth. The combined gravitational influence of the earth and the sun make its motion relatively complex. Not only does its orbital speed vary, but the tilt of its orbital plane relative to that of the earth round the sun can vary by up to about six degrees either way. Hence all lunar phenomena seem much less regular than those involving the apparent motion of the sun alone.

Even the nearest stars are vast distances away compared to the radius of the earth's orbit. Over a period of many years their relative positions as seen by an observer on the earth therefore change very little. As a result they may be taken as an almost fixed reference system against which phenomena within the solar system can be measured. By the beginning of the Christian era astronomers in both East and West had come to think of the stars as fixed on the inner surface of a vast rotating celestial sphere with the human observer at (or very near) its centre. The points where the sphere's imaginary axis passes through it are the north and south celestial poles. In reality the apparent celestial axis is simply the projection outwards into space of the earth's axis. Likewise the celestial equator, a great circle midway between the poles, is the outwards projection of the earth's equator. Greek astronomers, who believed in a spherical earth, were aware of these correspondences, but since Chinese astronomers lived in a flat-earth universe the poles and the equator remained solely celestial concepts. A second great circle, the ecliptic, is inclined to the celestial equator at about 23 degrees. This circle is the apparent annual path of the sun around the celestial sphere against the background of the stars. In reality it is simply the outward projection of the plane of the earth's orbit around the sun.

Further technical details of the phenomena just outlined can be found in standard texts such as Smart (1979). For our present purpose we are only concerned with the apparent motions of the celestial bodies as seen by an observer on the earth. In common with the majority of pre-Copernican inhabitants of the earth, the ancient Chinese believed that they lived on a stationary body, and they therefore took the apparent motions of the sun, moon, stars and planets as real motions. Since the coming of artificial lighting even educated people are rarely as familiar with the changing appearance of the night sky as their ancestors would have been. Some necessary reminders are therefore given in the course of the following discussion so far as space will allow. The best detailed introduction to this topic is given in the opening chapters of Kuhn (1979).

Days and day cycles

The observer assumed in the *Zhou bi* is in a latitude close to that of the Yellow River basin, around 35 degrees north. For such an observer the earth's daily rotation causes the sun to appear to rise over his eastern horizon, climb to its highest position due south of him at the moment of noon, and then sink back over his western horizon. Because of the tilt of the earth's axis relative to its orbital plane (which is the plane of the ecliptic), the time the sun spends above the horizon and the maximum altitude it attains both vary markedly in the course of a year as the earth orbits the sun.

No human society living on the surface of our planet can avoid using this repeating cycle of light and darkness as its basic unit of time reckoning. The *Zhou bi* is no exception when it states that 'daylight and night make one day' (#K8).[9] There have however been differences of practice as to when the division between days was placed. While the current civil practice in the majority of countries begins a new day at midnight, the religious calendars of Judaism and Islam make the division at dusk. Whatever the earliest practice in China may have been, by the end of the first millennium BC the convention of starting a new day at midnight was firmly established.

The *Zhou bi* does not raise the question of whether or not the day is of constant

[9] As noted above, the reference here is to section K of the *Zhou bi*, paragraph 8, according to my own division of the text.

length. The modern civil day is constant by definition, since it is taken to be precisely 24 hours, and each of these hours contains 3600 seconds defined by a standard atomic clock. A day defined by the cycle of the sun's apparent movement is by no means as simple. The combination of the earth's rotation on its axis with its orbital motion round the sun means that the time interval between successive noons as measured by a clock can vary by up to twenty minutes during the course of the year. For this reason time told by a sundial will not always be close to civil time as marked by a clock or watch. This fact will however prove to be of only minor importance in deciding how close the *Zhou bi*'s description of the phenomena comes to reality.

In the West, the artificial seven-day cycle of the week has long played an important role in structuring civil and religious time. In ancient China a ten-day period, the *xun* 旬, played an analogous role from at least as far back as the Shang dynasty. Each day was named using one of ten characters known as the *tian gan* 天干 'heavenly stems'. There is no consensus amongst scholars as to the original significance of these characters. By systematic pairing of the ten stems with another set of twelve cyclical characters (the *di zhi* 地支 'earthly branches') a longer cycle of sixty day-names was generated: see table 1. This sexagenary cycle was used for civil dating independent of months and years, and seems to have run unbroken up to the present from at least as far back as the beginning of the first millennium BC. During the Han dynasty it became customary to use the *gan zhi* 干支 cycle of sixty character pairs to designate a cycle of sixty years in addition to its continuing use for naming days.

Table 1. Stems and branches

	Stems	Branches
1	甲 *jia*	子 *zi*
2	乙 *yi*	丑 *chou*
3	丙 *bing*	寅 *yin*
4	丁 *ding*	卯 *mao*
5	戊 *wu*	辰 *chen*
6	己 *ji*	巳 *si*
7	庚 *geng*	午 *wu*
8	辛 *xin*	未 *wei*
9	壬 *ren*	申 *shen*
10	癸 *gui*	酉 *you*
11		戌 *xu*
12		亥 *hai*

The first decade of the sexagenary cycle begins with the stem–branch pair *jiazi* 甲子 as #1, and ends with *guiyou* 癸酉 as #10. The next decade begins with #11, *jiaxu* 甲戌 and continues with an offset of two in the stems relative to the branches. This process is continued, the offset increasing by two each decade, until we reach *guihai* 癸亥 #60, and the cycle then repeats. All one needs to do to convert stem–branch pairs to sexagenary numbers is therefore to find the offset (remembering to count forwards through the cycle of branches until one comes level with the stem) and divide by two to find the tens, and then take the units from the number of the stem. For an example, the reader may like to verify that *bingchen* 丙辰 corresponds to #53.

The lunar cycle

There is no hard evidence to decide the question, but it seems probable that long before the sixty-day cycle came into use the lunar cycle already served as a medium term time unit. In the earliest records we have (Shang oracle bones of the late second millennium BC) both units are already used in parallel, as they were to be ever after. Before entering into calendrical matters, it may be helpful to give a brief reminder of the appearance of the lunar cycle for an observer on the earth.

Following a lunation

Figure 1 indicates the physical basis of the varying appearance of the moon during the course of a complete lunar cycle, or lunation. When the moon is at A, its illuminated side is turned away from the earth, so that the observer cannot see it. This is the moment of conjunction, or 'new moon'. The Chinese term is *shuo* 朔 'dark'. Since the moon's orbit is in general not in the same plane as the earth's orbit round the sun, the moon is rarely exactly aligned between the sun and an observer on the earth; when this does happen the result is an eclipse of the sun in that observer's locality. If the moon was visible at conjunction, it would be very close to the sun in the sky, and would rise and set at almost the same moment as the sun. On the day when the moon is at B an observer on the earth sees a half-illuminated moon. This phenomenon, is called *xian* 弦 as are crescents in general. The sightlines to the sun and the moon are at right angles. As a result the moon appears to lag about a quarter of a day behind the sun in the sky. It rises near noon, reaches its highest point ('culminates') close to sunset, and sets about six hours after the sun. At C the moon is opposite the sun, and its visible side is fully illuminated. The moon now rises at about the time of sunset, and culminates close to midnight. This is the time of full moon, *wang* 望. The rare case of an exact sun–earth–moon alignment at full moon leads to an eclipse of the moon, as that body moves through the shadow-cone cast by the earth. Unlike an eclipse of the sun, an eclipse of the moon is seen by everyone who can see the moon at that time. At D the moon once more appears half illuminated, and is once more

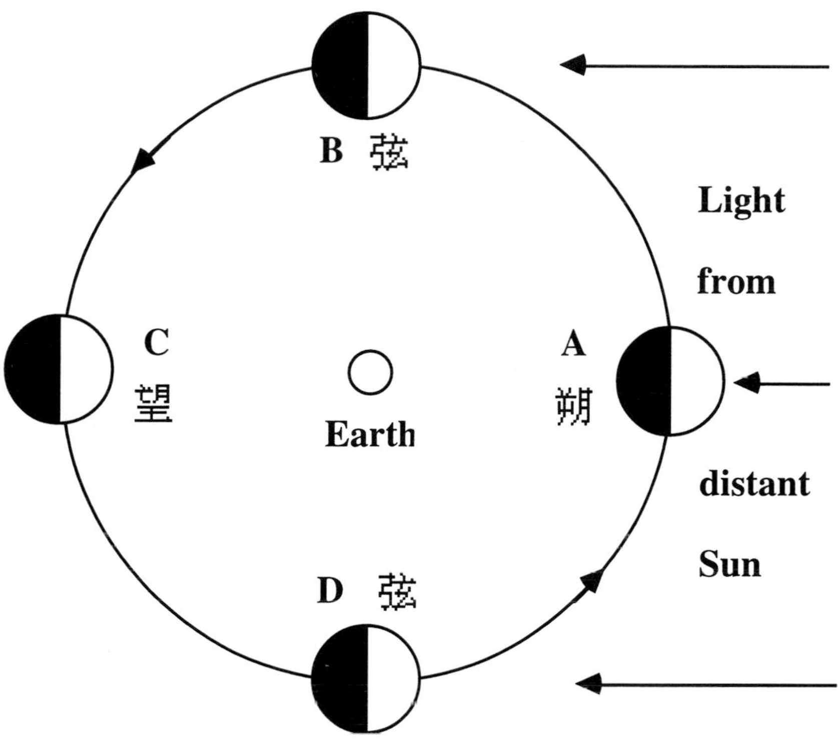

Figure 1. Following a lunation.

designated as *xian* 弦, but now lags three-quarters of a day behind the sun. It therefore rises about six hours before a subsequent sunrise, and culminates roughly at dawn. Finally it returns to position A, and the next new moon occurs.

As I have already mentioned, the moon's orbit is subject to complex variations. The result of this is that the length of a lunation is variable. It can be as short as 29 1/4 days, or as long as 29 3/4 days. The mean value is 29.5306 days, correct to six significant figures.[10]

Defining a month

As with the day, it is necessary to make an arbitrary decision about what instant in the repeating lunar cycle is to be counted as the starting point for time reckoning. By the end of the first millennium BC, Chinese astronomers had adopted the moment of

[10] This quantity is technically called a mean synodic month. To avoid confusion with civil months, which contain whole numbers of days, this term is avoided in the rest of the discussion, and the term 'lunation' is substituted.

conjunction for this purpose. Now for a given observer, the moment when the moon is most closely aligned with the sun can occur at any time of day, so that the first part of a day might fall in a different lunation from the part after conjunction.

Such a situation would be inconvenient for dating purposes. We thus meet for the first time the necessary distinction between the lunation and the lunar month. In Chinese usage, the lunar month is a period consisting of a whole number of days, whose first day is in principle the day during which conjunction occurs; in practice, as we shall see, various compromises and approximations are made. The lunar month is simply designated *yue* 月 'moon', and certain days in it are named after the lunar phenomena which fall within them, so that, for example, the first day, when conjunction occurs, is named *shuo* 朔 'dark', presumably because the moon is then invisible, with its illuminated side turned away from the earth. The months of the modern Gregorian calendar have no relation to the lunar phases, although their length of about 30 days makes it plain that European months were once 'moons' like their Chinese counterparts.

As will become clear later, by the beginning of the imperial age the decision as to when a new lunar month began was made by calculation rather than by observation. It is in any case impossible to observe the moment of conjunction by any simple means, except in the very rare case when the moon is precisely enough aligned with the sun to cause a solar eclipse. It is probable that at some stage in Chinese history the decision to begin a new lunar month depended on observation rather than on calculation.[11]

We cannot tell how this was done. It would have been possible to note the day of full moon, and guess that the next month would begin fifteen days later. More plausibly, observation could have been made closer to the end of the month. Typically there are about three days near conjunction when the moon is not visible at all. On the preceding day a thin crescent is still just visible rising over the eastern horizon a little before the sun appears. The next day is ideally the last day of the month, and is designated *hui* 晦 'dark of the moon'. Then follows the day of conjunction, *shuo*, the first day of the next month. The second day of the month, which has no special name, is still moonless. Ideally the evening of the third day should see the first thin crescent of the moon following the sun down over the western horizon at dusk. This is the day of *fei* 朏 'emergence of the moon'.[12]

In some calendars this day of first sighting of the new moon was the first day of the month: the Islamic fast month of Ramadan must still begin and end by direct observation of the first crescent. Perhaps at some early stage the first day of the month in China may also have been *fei* rather than *shuo*. But if one is aiming to start the

[11] A possible consequence of the reduction of conjunction prediction to rule may be seen in the *Analects* (*Lun yu* 論語) when (*c*. 500 BC) Confucius rebukes a disciple who wishes to discontinue the old custom of sacrificing a sheep to announce the first day of the lunar month in the ancestral temple of the Dukes of Lu: *Analects* III.17.

[12] The character is composed of graphs with the meanings 'moon' and 'emerge'.

month with a conjunction, it might seem reasonable enough to observe the last crescent of the preceding month, and start the new month on the next day but one. In reality however the visibility of the first and last crescents is highly dependent on the precise position of the moon relative to the sun as they near the horizon, as well as on the season of the year and the observer's latitude, both of which affect the angle at which the sun and moon move through the horizon plane. Simple crescent observation and day counting will not therefore always fix the conjunction day correctly. This is however of little practical importance, since one is unlikely to be much more than a day out. As a result the first day of the month so defined will still be moonless, although the first crescent may be a little earlier or later than it ideally should be. Except when direct earth–moon–sun alignment produces a solar eclipse, conjunction cannot be verified directly by the naked eye, so the error will not become obvious. During the Han period Chinese astronomers were thus able to work with a mean lunation which predicted conjunctions at fixed intervals without blatant contradiction from observation.

The year

Like our year, the Chinese civil year began at the instant that a new day began. It therefore always consisted of a whole number of days. Unlike ours, it also began when a new lunar month began, and therefore always consisted of a whole number of lunar months, normally twelve. As we shall see, Chinese calendrical thinking has a tendency to build up nests of cycles, so that the initial day of each one is also the initial day of each of the shorter cycles it contains. One of the paradigmatic problems of Chinese calendrical theory is the attempt to find the 'Grand Origin Point' *shang yuan* 上元 at which all cycles begin together.[13] A New Year's Day that was not also a new moon would have seemed as unnatural as Easter Day on a Tuesday rather than a Sunday would seem to most modern Christians.

A series of twelve lunar months will generally turn out to be about eleven days shorter than the annual cycle of the seasons as represented, for example, by the return of the winter solstice.[14] Left to itself, the twelve-month cycle will therefore get more

[13] See paragraph #K4.

[14] See the detailed discussion of intercalation below. To avoid confusion, it is important to distinguish the concept of the Chinese civil year (which is a whole number of lunar months, each consisting of a whole number of days) from the so-called 'tropical year', which is the interval at which solstices recur (strictly speaking the cycle is from one spring equinox to the next). A modern value for the tropical year is 365.2422 mean solar days. As we shall see shortly, we must distinguish the tropical year from the sidereal year of 365.2564 days, after which the sun returns to its original position relative to the stars. But since Han astronomers did not yet make this distinction I use the term 'solar cycle' to refer to the period (365 1/4 days for the *Zhou bi*) during which it was believed both solstices and solar position would recur.

and more ahead of the seasons, so that over a few years the first month will move through spring, winter, autumn, and summer in turn. In some cultures this has not been seen as an important point. Islam uses such a pure lunar calendar, and there are no signs that either civil administration or the needs of agriculture in Islamic countries have suffered in consequence. And although the ancient Egyptian calendar was not lunar, its use of a fixed year-length of 365 days caused a slower but steady slippage without obvious signs of strain for the Egyptian peasant. In China, however, the decision to try to keep the civil year in step with the seasons seems to have been made from an early period, certainly as far back as the Shang dynasty. It seems likely that this choice was associated in some way with the importance of seasonal ritual activities performed by Chinese monarchs as part of their task of maintaining harmony between the order of human society and the overall order of the cosmos.

The annual cycle of nature as we perceive it is made up of many interlocking cycles, involving changes in weather, the growth cycles of plants, the breeding cycles of animals, and the changing positions and visibility of the sun and stars. There is no obvious reason why one type of cycle should be singled out from the rest, and indeed there is some evidence that early Chinese calendrical thought made use of the full range of reference cycles in defining the phenomena that ideally belong in a given month.[15] Human activity, whether defined as ritual or practical in modern terms, is just one component part of the cosmic pattern. But by the end of the first millennium BC it is clear that the primary points of reference for calendrical purposes were certain types of astronomical events. These were presumably favoured in part for their regularity and relative ease of observation, but we should not miss the ideological significance of the decisive shift towards the primacy of the heavens over mere earthly phenomena. There is certainly a link here with the cosmological underpinnings of the unified empire in which the emperor was to his subjects as heaven was to earth. The *Zhou bi* discusses such astronomical reference points in some detail, as we shall see.

Astronomical markers for seasons

As the year passes, the sun goes through an obvious cycle of behaviour. For a northern hemisphere observer, at the winter solstice (*dong zhi* 冬至 'winter extreme') in midwinter it climbs above the south-eastern horizon relatively late. At noon it is still fairly low in the sky due south, and it sets roughly south-west. As the year progresses, it rises earlier and sets later, and the points at which it crosses the horizon move northwards.

15 See for example the monthly phenomena listed in the pre-Han annuaries *Xia xiao zheng* 夏小正 (*Congshu jicheng* edn) and the *Yue ling* 月令 chapters of the *Lü shi chun qiu* 吕氏春秋 (*Sibu congkan* edn). In the first, astronomical phenomena are only one category of seasonal indicators amongst others, including the weather, and the behaviour of plants, birds, fishes and animals. In the second, dating from about 239 BC, the astronomical indicators have won pride of place and are listed first, although the others are also mentioned.

The noon sun is a little higher in the south each day. When the summer solstice (*xia zhi* 夏至 'summer extreme') is reached the day is at its longest; the sun rises in the north-east and sets in the north-west. At noon it reaches its greatest altitude of the year. After the summer solstice the movement reverses, and after another winter solstice the cycle repeats.

Meanwhile a slightly less obvious cycle is taking place in the relation of the sun and the stars. If we could see the stars at the same time as the sun, we would notice that the celestial sphere on which the stars may be imagined to lie rotates once a day, bearing the sun with it. The axis about which this apparent rotation occurs passes through the observer at an angle to the horizon equal to his latitude. The points where the axis passes through the celestial sphere are the north and south celestial poles, and half-way between them is the celestial equator. However, the sun does not keep to a fixed position amongst the constellations. As the earth orbits round the sun, the sun is naturally seen against a changing stellar background. As a result the sun appears to trace out a circle round the celestial sphere in the course of a year: this is the ecliptic.

While this apparent solar motion cannot be directly observed, it is obvious that in consequence the stars visible at (say) midnight will also change slightly from day to day. The stars are in fact taking a little less time to return to their position relative to the observer than the interval between successive noons or successive midnights. There is thus a typical midwinter night sky, most of whose stars will be below the horizon during a midsummer night, and *vice versa*. Because of the phenomenon of precession the appearance of the night sky at a given season also changes slowly, passing through a complete cycle in about 23 000 years. During the period with which we are concerned Chinese astronomers were unaware of this fact, and assumed that the solar and stellar cycles had exactly the same period, so that (for example) the winter solstice sun was always at the same position amongst the stars.[16]

As a result, early Chinese calendrical specialists were free to observe whatever they felt to be the most significant or convenient component of what was for them a single cycle of astronomical phenomena. From observation of one set of phenomena, the progress of all others could be deduced. I will now briefly discuss three of the main methods used by the *Zhou bi* for this purpose.

Noon shadows and the solstices One obvious indicator of the progress of the solar cycle is the sun's altitude above the southern horizon at noon, when it reaches its highest point on a given day. Midwinter's day has the lowest noon sun, and midsummer's day has the highest. These changes can be followed by the use of a simple vertical pole gnomon, whose shadow shortens and lengthens as the sun rises and falls in

[16] For a discussion of the difficulties that ensued when it became evident that the conventional winter solstice position was in error, see Maeyama (1975–1976).

relation to the horizon. Midwinter's day is marked by the longest noon shadow, and midsummer's day by the shortest. It therefore seems a simple matter to establish the length of the solar cycle: all one has to do is to count the days from (say) one midwinter's day as determined by the gnomon to the next.

In practice the task is not so easy. In the first place the sun's noon altitude changes quite slowly near its maximum and minimum, and it is therefore hard to be sure on exactly which days these occur. Secondly, the length of the cycle revealed by this method is not quite constant from year to year. Even assuming perfect accuracy in locating the critical days, while most cycles will turn out to be 365 days long, some will turn out to be 366 days long. Over a long enough period the average will turn out to be something like 365 1/4 days. This point is made in the *Zhou bi*, paragraphs #K9 to #K11. Clearly the real solar cycle is about something other than what the sun happens to be doing at noon.

In the language of spherical astronomy, what is happening is that in this period the sun is moving once round a great circle on the celestial sphere against the background of the stars. Its north–south movement is caused by the fact that this circle, the ecliptic, is inclined to the celestial equator by about 23.5 degrees. Its furthest point north is at the instant of summer solstice, and the furthest point south is at the instant of winter solstice. The period of this cycle is known as the tropical year; to seven significant figures it is 365.2422 days. The phenomena of sunrise, noon and sunset are governed by the apparent daily rotation of the celestial sphere relative to the observer, and it is therefore clear that the actual moments of solstice will not bear any fixed relation to the moment of noon, so that the solar cycle will not be a whole number of days.

As well as the two solstices there are two other obvious points of importance on the sun's annual circuit of the stars. These are the spring and autumn equinoxes (*chun fen* 春分 'division of spring' and *qiu fen* 秋分 'division of autumn'), when the sun passes through the celestial equator. At these times the sun rises exactly due east and sets exactly due west; it spends twelve hours above the horizon and twelve hours below. During the Han dynasty it was assumed that the sun took equal amounts of time to move through each of the four quarters into which the its path was divided by the solstices and equinoxes, so that the seasons were of equal length.

Each quarter was further divided into six equal divisions, so that the year was divided into 24 periods. These are the 24 *qi* 氣, a term which is applied equally to each period and to the instant of its inception. A list of them appears in paragraph #H2 of the *Zhou bi*. From the calendrical point of view, the chief function of this system is to make it easy to check whether the lunar months bear the intended relation to the seasons. Since there are 24 *qi*, and a normal year contains twelve lunar months, it is easy to see that each lunar month will tend to contain two *qi* within itself. In a sense, each pair of *qi* may be regarded as a 'solar month', but it must be remembered that these 'months' do not have a whole number of days in them, and that the instant of *qi* inception can fall at any time of day. Out of the 24 *qi* only the solstices and equinoxes are defined by any obvious feature of the sun's behaviour.

Once the length of the cycle is known the instant of each *qi* inception can be found as soon as one of them has been fixed. This reference point is usually taken to be the winter solstice. Subsequent *qi* inceptions may be located by counting off time intervals equal to one twenty-fourth of the solar cycle. The list of *qi* given in the *Zhou bi* is accompanied by a table of noon shadows for a standard gnomon (paragraph #H2), but most of these are theoretical constructs based on linear interpolation between solstitial values, and bear no direct relation to observation.

Solar right ascension and longitude The origins of the *qi* system are linked to observations of the sun's noon altitude through the length of shadow cast, and hence (in modern terms) to its varying declination, or distance north or south of the celestial equator. A different reference system is used to deal with what we would now describe as the sun's movement in right ascension as it travels from west to east round the celestial sphere.[17] This is the system of the 28 lodges, *xiu* 宿. The origin of the system is unclear, but it was well established by the end of the first millennium BC. If we allow ourselves to think in terms of the celestial sphere (which the originators of the system almost certainly did not) the lodges divide the sphere into 28 slices of right ascension like the segments of an orange. This is however a very asymmetrical orange, since the segments vary in width from about 1.5 degrees to about 30 degrees.[18]

Ideally, the beginning of each lodge is marked by a reference star, although in reality the correspondence between the positions of stars and the theoretical extent of the lodges is not exact. The names of the lodges, with their reference stars, are tabulated in table 2. Recent archaeological evidence combined with an examination of literary sources has revealed the existence of an 'old system' of lodges differing in important respects from its successor, which seems to have been created in the great

[17] In terrestrial terms declination corresponds to latitude and right ascension to longitude. As terrestrial longitude is measured from the meridian of Greenwich, right ascension is measured from the position of the sun at the spring equinox. Confusingly, celestial longitude and latitude are not the same things as right ascension and declination, but are used in a different reference system, based on the ecliptic rather than the celestial equator.

[18] The actual Chinese unit used is the *du* 度 'graduation'; it is close to the Western degree, since 360° = 365 1/4 *du*. During the Western Han it was assumed that the sun moved through the system of the *xiu* at a constant 1 *du* per day, so that it completed its circuit in 365 1/4 days.

Table 2. New and old lodge systems

	Name	Reference star and width in new system	(*du*)	Reference star and width in old system	(*du*)
1	Jue 角 'Horn'	α Virginis	12	α Virginis	12
2	Kang 亢 'Gullet'	κ Virginis	9	κ Virginis	11
3	Di 氐 'Base'	α^2 Librae	15	α^2 Librae	17
4	Fang 房 'Chamber'	π Scorpii	5	π Scorpii	7
5	Xin 心 'Heart'	σ Scorpii	5	α Scorpii	11
6	Wei 尾 'Tail'	μ^1 Scorpii	18	λ Scorpii	9
7	Ji 箕 'Winnower'	γ Sagittarii	11 1/4	γ Sagittarii	10
8	Dou 斗 'Dipper'	φ Sagittarii	26	φ or σ Sagittarii	22
9	Niu 牛 'Ox'	β Capricorni	8	α^2 Capricorni	9
10	Nu 女 'Woman'	ε Aquarii	12	ε Aquarii	10
11	Xu 虛 'Barrens'	β Aquarii	10	α Equulei	14
12	Wei 危 'Rooftop'	α Aquarii	17	θ Pegasi	9
13	Shi 室 'House'	α Pegasi	16	η Pegasi	20
14	Bi 壁 'Wall'	γ Pegasi	9	α Andromedae	15
15	Kuei 奎 'Straddler'	η Andromedae	16	β Andromedae	11
16	Lou 婁 'Harvester'	β Arietis	12	β Arietis	15
17	Wei 胃 'Stomach'	41 Arietis	14	β Persei	11
18	Mao 昴 'Mane'	η Tauri	11	17 Tauri	15
19	Bi 畢 'Net'	ε Tauri	16	α Tauri	15
20	Zui 觜 'Beak'	λ^1 Orionis	2	λ^1 Orionis	6
21	Shen 參 'Triaster'	ζ Orionis	9	α Orionis	9
22	Jing 井 'Well'	μ Geminorum	33	γ Geminorum	29
23	Gui 鬼 'Ghost'	θ Cancri	4	θ Cancri	5
24	Liu 柳 'Willow'	δ Hydrae	15	δ Hydrae	18
25	Xing 星 'Star'	α Hydrae	7	ι Hydrae	13
26	Zhang 張 'Spread'	μ Hydrae	18	μ Hydrae	13
27	Yi 翼 'Wing'	α Crateris	18	γ Crateris	13
28	Zhen 軫 'Axletree'	γ Corvi	17	γ Corvi	16

calendrical reform of 104 BC.[19] For comparison the old system has been included in the tabulation. A celestial object is said to lie in a particular *xiu* if it crosses the meridian after the relevant reference star, and before the reference star of the next *xiu*. It is irrelevant whether the object comes close to the asterism to which the reference

[19] See the discussion of Wang and Liu (1989).

star belongs. The reference stars do not seem to have been chosen to lie close to either the ecliptic or the celestial equator at any epoch, and they certainly did not need to do so to serve their purpose.

The data for the 'new' system are taken from Needham (1959) table 24. Those for the 'old' system are from Wang and Liu (1989), using the lodge widths to the nearest whole *du* from the excavated 'lodge disc' there described. In some cases identifications of reference stars in the old system are open to discussion. For lodges 25, 26 and 27 Wang and Liu label stars as L, M and r of the relevant asterisms, evidently through typographical error. Lodge names are translated as in Schafer (1977) table 2.

Centred stars It is not possible to observe the position of the sun amongst the lodges directly, since its brightness makes it impossible to check which reference stars cross the meridian before and after it. Any estimate of its position must necessarily involve theoretical inference as well as observation. It is not surprising, therefore, that what appear to be our earliest texts referring to the seasonal changes in the heavens do not refer to the sun in direct relation to the stars. The *Xia xiao zheng* 夏小正 'Lesser Annuary of the Xia' mentions fifteen stellar phenomena linked to eight of its twelve months, including when certain asterisms first become visible in the night sky or are last seen, the dusk or dawn direction of the Dipper handle, and in three instances references to a star being *zhong* 中 'centred' at dawn or dusk.[20] If later usage is a guide, this expression refers to the moment when a star crosses the observer's meridian due south. Unfortunately the origin and date of this text are highly problematic, although it certainly resembles no text of the Han period.

It was the dusk and dawn 'centring' of stars that was to become the most important directly observable stellar seasonal phenomenon in early Chinese astronomy. There is no doubt that the four important stars mentioned as seasonal markers in the astronomical foundation charter in the *Yao Dian* 堯典 (part of the Book of Documents, perhaps composed in the fourth century BC) are to be interpreted in this way. Despite heroic attempts from the time of Gaubil in the eighteenth century, efforts to date this text from the stars it mentions seem misdirected, since they discount the large number of uncontrolled assumptions necessary to begin a quantitative analysis of such data.[21] The earliest datable listing of centred stars is given in the *Yue ling* 月令 'Monthly Ordinances' chapters of the *Lü shi chun qiu* 呂氏春秋, completed in 239 BC. The fact

[20] See this short text in the *Congshu jicheng* edn.

[21] Thus we have to identify the stars concerned (which are probably asterisms of finite breadth), choose exactly what times of the year are referred to, guess a latitude for the observer, and guess what is to count as 'dusk' and 'dawn'. Some of the difficulties involved in such calculations were (so far as I can tell) first pointed out by Ptolemy around AD 150: see *Almagest* VIII.6.H203, pp. 416–17 in Toomer (1984). On this topic see also Yabuuchi (1969), 267. Some of the attempts to date this and related material are discussed in Needham (1959), 245–7.

that this listing also includes solar positions amongst the lodges warns us that the data given come from a time when astronomical theory may be as important an influence as observation.

The first full listing of dusk and dawn centred stars in the context of an integrated calendrical scheme is given in the system of AD 85 in *Hou Han shu* 後漢書, *zhi* 志 3, 3077–8. The seasonal reference points are the twenty-four *qi* rather than the vague months of the *Lü shi chun qiu.* The data are given together with the lengths of day and night as well as the position of the sun amongst the lodges. Once again observation and theory are inextricably entwined.

Observations of centred stars are for practical purposes an easy way to keep a running check on whether or not the calendar is in close step with the heavens. By contrast, solstice observations require careful measurement and can only be done twice a year. Perhaps because of its stress on the role of the gnomon for shadow observation, the *Zhou bi* does not, however, discuss the use of centred stars as seasonal indicators, although it does discuss the meridian transits of stars at other times.

The structure of an astronomical system

How to forget astronomy

It is not a difficult matter to run a luni-solar calendar of the Chinese type if all one is required to do is to say what the date is today. The sixty-day cycle is a matter of simple counting, and with a little boldness in extrapolating from observation of the last crescent to estimate the day of conjunction, it is simple to define the day of the month. If at the end of a civil year of twelve lunar months it becomes clear that the year is slipping seriously out of step with the appropriate phenomena amongst plants, animals, the weather and the heavenly bodies, one just adds an extra 'intercalary' month to that particular year to allow nature to catch up before reckoning the first month of the New Year.

The problem about this is that prediction is not part of the deal. You cannot, for instance, tell in advance what day of the sixty-day cycle will be the next New Year's day, although a rough guess is of course possible. This sets limits on the planning of activities for the year ahead, especially when state ritual activity appropriate for a particular day is involved. At a less exalted social level, it is impossible to predict lucky and unlucky days very far in advance. It was this problem that Chinese calendrical astronomy set out to solve by creating the mathematical systems known as *li* 曆, a term conventionally but misleadingly translated as 'calendar' (and to be distinguished from the similarly romanised length unit *li* 里). In reality these were astronomical systems designed to make astronomical observation superfluous by providing rules for predicting all important celestial phenomena. Such systems were commonplace by the end of the first millennium BC. Their construction demanded a measure of approximation and compromise related inversely to the mathematical and astronomical resources available to those who created them. In the Western Han, when full descriptions of such systems first became available, the degree of approximation was often relatively large.

As the discussion proceeds, it will become clear that the *Zhou bi* does not contain a complete description of a *li*. It does however give the basic constants which form an important part of all *li* of a particular type, the 'quarter-remainder' or *si fen* 四分 group. It is these constants that will be used for illustration in what follows.

The month and the day: long and short

On a naive view, to predict on what day of the sixty-day cycle a future month begins, it is necessary to find on what day the appropriate conjunction of sun and moon occurs. This is not an easy matter since, as already mentioned, the behaviour of the sun–moon–earth system is complex. The intervals between successive conjunctions can vary between about 29.25 days and 29.75 days. Further, it must be remembered that the days used here are mean days. Days as actually observed with reference to the sun vary in length by about twenty minutes in the course of a year, complicating matters further.

All this is however an anachronism if we are to look at matters from the point of view of an ancient Chinese calendrical astronomer. All he had to do to be hailed as successful was *not to be seen to be wrong*. Since the conjunction cannot be observed directly, he would be safe so long as he was not much more than a day out, and thus avoided the risk of the crescent moon being visible on the alleged conjunction day. It will be recalled that there are generally moonless days preceding and following conjunction. To achieve this degree of success, it is only necessary to predict the 'mean lunation', that is, to find the days on which conjunctions would occur if they were spaced at constant intervals equal to the average of the variable spacing of real lunations.

The modern value of the length of the mean lunation is 29.53059 mean days, correct to seven significant figures. Since this is close to 29 1/2 days, early Chinese astronomers were able to follow the mean lunation quite closely for some time by simply alternating 'small months' *xiao yue* 小月 of twenty-nine days with 'large months' *da yue* 大月 of thirty days. If the alternation is unbroken, one is in effect neglecting 0.0306 days each month, and after about thirty months the accumulated deficit will amount to a whole day. To take account of this it is necessary to allow two large months in succession to occur at predetermined intervals. Precisely when this is done depends on the value for the length of the mean lunation used by the astronomical system in question. As will be seen in the next section, the *Zhou bi* uses a mean lunation length of 29 499/940 days (29.53085 days to seven significant figures). There is however no indication of when successive large months are to be inserted.

The month and the year: intercalation

The interval between successive occurrences of the same solstice changes so little that in the present context it may be taken as constant at 365.2422 days, correct to seven significant figures. Chinese astronomical systems used a variety of values for this quantity. The *Zhou bi* is committed to the value of 365 1/4 days for the solar cycle,[22] and a number of different *li* have this feature in common. They are therefore said to be of the 'quarter-[remainder]' *si fen* 四分 type.

It will be recalled that twelve lunar months give a civil year about eleven days shorter than the solar cycle, so that it is necessary to add an extra 'intercalary' month every three years or so to keep in step with the seasons. If precise values are assigned for the mean lunar and solar cycles, it is possible to calculate the average frequency with which intercalations must be made over the long term. The *Zhou bi* implies in paragraphs #K9 to #K11 that the solar cycle length is determined by observations of successive winter solstices over several years.

The mean length of the lunar month is not however fixed independently. Instead, the *Zhou bi* gives logical priority to a statement of how many lunar months there should be in a certain number of civil years: see #I2 and #K15, #K16. The implication is that over a long enough period it should be possible to find a precise common multiple of the solar cycle and the mean lunar cycle. Given a value for the solar cycle, the mean lunar cycle may then be derived. The equivalence used in paragraph #I2 is:

19 solar cycles = 235 mean lunar cycles.

Nineteen solar cycles of 365 1/4 days do not amount to a whole number of days, and hence this period cannot contain integral numbers of civil years and civil months. Perhaps for that reason the direct calculation of the month length made in #K15 and #K16 is based on the number of mean lunar cycles in 76 years, for which we may say:

76 civil years
= 76 solar cycles
= 4 × 235 months
= 4 × 235 mean lunations.

[22] See #D6 and #K11.

Thus one mean lunation = $76 \times (365\ 1/4 \text{ days}) \div (4 \times 235)$, from which it follows that the mean length of the lunar cycle must be 29 499/940 days.

From the equivalences used here it is possible to determine the intercalation rule which lies behind them. If 76 years contain $4 \times 235 = 940$ months, then the number of intercalary months added in this period must be:

$$940 - (76 \times 12) = 28.$$

This may be stated in its simplest terms by saying that there must be seven intercalations in a nineteen-year period. The *Zhou bi* does not discuss how these extra months are to be distributed. Indeed, it does not make any explicit mention of intercalation at all, although paragraphs #I3 to #I7 refer to the 'small year', 'large year' and 'mean year'. The common term *run yue* 閏月 'intercalary month' does not occur in the text.

The system origin

On the basis of the data discussed so far, we can state the intervals at which solstices and mean conjunctions should follow one another. But to determine when a future solstice or conjunction should fall we need to have a fixed point from which to start our reckoning. We would have all we needed if the instants of any solstice and any conjunction were specified. The two events could be centuries apart without creating any difficulties of principle. In practice, however, by the beginning of the Han dynasty it is clear that Chinese astronomers preferred to do things differently.

All early *li* for which we have any evidence specify that all reckonings are to begin from a single 'system origin' *li yuan* 曆元 , or epoch. At this point it is usually claimed that a number of important cycles began simultaneously. In the first place, a new day begins, so that the instant of system origin is midnight. This new day may begin the sixty-day cycle, or represent some cosmologically significant point within it. It is also usual to specify that there is a mean conjunction at midnight, so that the lunar cycle begins with the start of the new day. Thirdly some important instant of the solar cycle, perhaps the winter solstice, or at least some other *qi* inception, will also fall at midnight. If the winter solstice is chosen, the month beginning will be the eleventh of the so-called Xia 夏 [dynasty] count, which is still used in the traditional Chinese calendar today.

The choice of a system origin has a twofold significance. The first is a cosmic one: both the conditions at the system origin and the year in which they occur may be seen as having numerological and political importance. Secondly, the existence of a common origin for many cycles considerably simplifies calculations: how this works will be seen in the next section. It is striking that while the *Zhou bi* makes clear reference to the second point, it nowhere specifies a system origin. As a result it contains insufficient information to enable actual calendrical calculations to be made. The significance of this omission will be discussed later.

Resonance periods

It was not uncommon for the system origin of a *li* to be fixed some centuries earlier than the period during which the *li* was devised and put into practice. Since the *Zhou bi* does not touch on such questions, a full discussion of how a system origin was chosen would be out of place here. It has already been indicated that the choice might be strongly influenced by political and cosmological considerations. In addition, the combination of the system origin and the basic constants of the *li* in question had to produce an acceptable degree of success in predicting when solstices and mean conjunctions would fall during the period of currency of the given system.

It is easy to see that making such predictions could involve manipulation of quite large numbers. Thus suppose one wished to know the day of the month on which winter solstice fell 330 years after the system origin. We need, in effect, the remainder from the following division:

$$(330 \times 365\ 1/4) \div (29\ 499/940).$$

To perform this using ancient Chinese computational procedures, the top and bottom of this fraction must be converted to integral form, obtaining:

$$\{[330 \times (4 \times 365 + 1)] \times 940\} \div [(29 \times 940 + 499) \times 4]$$
$$= (453\ 202\ 200) \div (111\ 036)$$
$$= 4081 \text{ months, with remainder } 17.1 \text{ days.}$$

Fortunately such calculations are not essential, since a number of useful periodicities are produced by the choice of constants used in the *Zhou bi* and in several ancient *li*. Paragraph #K4 specifies these, without however stating the uses to which they might be put. A brief explanation is therefore given here. We may assume that we are using a system origin at which *qi* inception and conjunction coincide at a given time of a day of stated number in the sixty-day cycle. The conjunction specified is usually that which begins the first month of the civil year, so that the system origin falls on New Year's Day.

(1) 19 years: one *zhang* 章 (the 'Rule Cycle').[23]

[23] In the ancient West, this period is associated with the name of Meton of Athens (fl. 430 BC): see Neugebauer (1975) vol. 1, p. 4, and vol. 2 p. 622. After specifying each cycle I give the translation suggested in Sivin (1969). The names for the resonance periods given here are those most often used in

19 × 365 1/4 days = 235 × 29 499/940 days
and 235 = 12 × 19 + 7.

It is thus clear that if we label the year that begins with system origin as year #1, and insert seven intercalary months during the rest of the *zhang* period, then New Year's Day of year #20 will once more see exact coincidence of the specified *qi* inception and conjunction. Year #20 may then be relabelled as year #1 of the next *zhang*, during which the pattern of intercalation is precisely repeated.

(2) 76 years: one *bu* 蔀, equal to four *zhang* (the 'Obscuration Cycle').[24]

Since 19 × 365 1/4 days = 6939 3/4 days, it is clear that we need to wait four *zhang* until *qi* inception and conjunction coincide at the same time of day as at system origin.

(3) 1520 years: one *ji*, 紀 equal to 20 *bu* or 80 *zhang* (the 'Era Cycle').

A *bu* contains 27759 days. Since 27759 = 60 × 462 + 39 days, successive *bu* will not commence on the same days of the sixty-day cycle. 20 *bu* is the smallest multiple which will produce a number of days divisible by sixty. The commencement of each *ji* therefore repeats system origin conditions for *qi* inception, conjunction, time of day and day number.

This last 'resonance period' is the longest one likely to be of any practical use. One more is however known in *li* of the quarter-remainder type. It is:

(4) 4560 years: one *yuan* 元 , equal to 3 *ji*, 60 *bu*, or 240 *zhang* (the 'Epoch Cycle').

During the Han period it became usual to name years as well as days using the sixty-fold cycle. By providing a multiple of 60 years, the *yuan* ensures repetition of the year number as well as all other system origin conditions.

Finally, the *Zhou bi* mentions a resonance period which occurs nowhere else:

(5) 319 20 years: one *ji* 極 (note different character from that in 3 above), equal to 7 *yuan*.

Since no multiple of seven occurs in ancient Chinese time-reckoning, it is hard to see the function of this period. One possibility seems to be that some rumour of the seven-day week had reached China from the West, and that this period is designed to ensure the repetition of the day of the week as well as all other system origin conditions.[25]

quarter–remainder systems, as speciified in *Hou Han Shu, zhi* 3, 3058–9. The *Zhou bi* uses a variant set: see #K4.

[24] In ancient Greece linked with Callippus (fl. 330 BC): see Neugebauer (1975) vol. 2, 623–4.

[25] I owe to Jochi Shigeru (private communication) the suggestion of a purely Chinese origin for the

How can the use of these resonance periods simplify the calculation of conditions when, as in the example we are considering, we are 330 years after system origin? We may proceed as follows.

$$330 = 76 \times 4 + 26$$
$$\text{and } 26 = 19 + 7.$$

We are therefore commencing the eighth year of the second *zhang* of the fifth *bu* since the system began to run (it must be noted that the first *zhang*, *bu* etc. began at system origin). For our example, system origin had conjunction and winter solstice coinciding at midnight, and these conditions will have repeated at the beginning of the fifth *bu*. Each of the four *zhang* in a *bu* commences with the same coincidence but 3/4 day later each time. It will be recalled that the pattern of insertion of the seven intercalary months is the same in each *zhang* of nineteen years. Now since

$$7 \times 7/19 = 49/19 = 2\ 11/19$$

it is evident that during the first seven years of the *zhang* there will have been two intercalations. Therefore the total number of months between the beginning of the *zhang* and the eighth New Year's day will be:

$$7 \times 12 + 2 = 86.$$

This number of mean lunations is equivalent to

$$86 \times 29\ 499/940 \text{ days} = 2539\ 614/940 \text{ days.}$$

On the other hand, seven solar cycles is equivalent to:

$$7 \times 365\ 1/4 \text{ days} = 2556\ 3/4 \text{ days.}$$

Thus the solstice falls later than the conjunction of the New Year by:

$$2556\ 3/4 - 2539\ 614/940 \text{ days} = 17\ (705 - 614)/940 \text{ days}$$
$$= 17\ 91/940 \text{ days.}$$

factor of seven. In the earliest Chinese discussions of problems equivalent to modern indeterminate equations, remainders from an unknown number divided by three, five and seven are typically given. One of the most obvious practical applications of indeterminate problems was the determination of the *li yuan* 曆元 system origin from presently observed phenomena. The factor of five has been introduced in the *ji*, and the factor of three in the *yuan*. Could it be that the sole purpose of the final resonance period is to introduce the missing third factor of seven?

Clearly the solstice falls on the eighteenth day of the first month of the civil year. It is obvious that proper use of the system of resonance periods enables us to make all such calculations without computations involving unnecessarily large numbers.

Calendrical astronomy under the Han

I will now give an outline of changes in the official astronomical system under the Han dynasty. No more detail will be given than is necessary for discussion of the origins of the *Zhou bi.*

The Zhuan Xu li

The twenty-sixth regnal year of King Zheng 正 of the state of Qin 秦 ran from late 222 BC to late 221 BC. After years of ruthless warfare Qin had overcome all rivals and united China under a single imperial government. As part of the new dispensation, the king inaugurated a new astronomical system, the *Zhuan Xu li* 顓頊曆 'Zhuan Xu system'. Zhuan Xu was the grandson of the legendary Yellow Emperor of high antiquity, and it was from him that the royal house of Qin claimed descent. This system was used throughout the Qin dynasty. When the first Han emperor was proclaimed in 202 BC he continued to use much of the state machinery and ritual of his predecessors, including the *Zhuan Xu li,* which was not replaced until nearly a century later.

The *Zhuan Xu li* was a system of the quarter-remainder type described above. As has already been indicated, such systems differ amongst themselves only in their choice of *li yuan* 曆元 'system origin' or epoch. According to the account given by Chinese historians in the first century AD, at the time of the Qin unification there are said to have been six different astronomical systems in circulation *(Han shu* 21a, 973). Two of these, the *Huang Di li* 黃帝曆 and the *Zhuan Xu li* 顓頊曆, bore the names of semidivine rulers of mythical antiquity, the so-called Yellow Emperor and his grandson respectively. The *Xia li* 夏曆 , *Yin li* 殷曆 and *Zhou li* 周曆 were named after what were seen as the first three great dynasties to rule China, while the *Lu li* 魯曆 was named after the feudal state in which Confucius had lived. In only the last two cases does there seem any possibility of a real historical basis for the name. At the end of the first century BC the imperial library contained works said to describe all of these astronomical systems, amounting to 82 *juan* 卷 in total (*Han shu* 30, 1765–6). Only the titles of these books remain today, but references in other works enable us to say a good deal about ancient astronomical systems in general; in the case of the *Zhuan Xu li* and to a lesser extent the *Yin li* we can fill in the details of their special features.

From 104 BC onwards the standard histories have preserved more or less complete details of the successive astronomical systems in official use.

The *Zhuan Xu li* appears to have specified the following initial conditions:[26]

> Year: 26th regnal year of Duke Xian 獻 of Qin, i.e. late 367 BC to late 366 BC. This year is designated as 51 in the 60-year cycle. During this year the precise instant of inception of the fourth *qi*, *li chun* 立春, falls at midnight and coincides with the instant of conjunction of sun and moon, which are then in the fifth *du* 度 of the lodge House; the day then beginning is 51 of the 60-day cycle, and the month is the *zheng yue* 正月, the first of the Xia month-count.[27]

The unit *du* 度 used here results from dividing the circumference of a circle into 365 1/4 parts: the link with the length of the year is obvious. A check using modern astronomical tables shows that the astronomical events referred to all fell within the first twelve hours of 366 BC February 9, according to local time in the central Yellow River basin. It is however highly unlikely that the astronomers who devised the *Zhuan Xu li* nearly 150 years later had any reliable observational data for the day in question. The choice of a system origin would be more likely to relate in part to the accuracy of the system's predictions at the time of its creation and in part to the neatness of supposed coincidence of events under system origin conditions. Other quarter-day systems using different *li yuan* might place their initial *qi*/conjunction/midnight coincidence in another year altogether. In practice, however this discrepancy would do no more than shift the predicted instants of *qi* inceptions and conjunctions by various fractions of a day relative to the *Zhuan Xu li* reckoning.

The Tai Chu li and the San Tong li

The first emperor of the Han dynasty began his rule without the panoply of a new cosmic dispensation. Near the end of the second century BC the Emperor Wu 武 decided to repair the omission by a comprehensive revision of state ritual and symbolism, which naturally included a new astronomical system, which was to be known as the *Tai chu li* 太初曆 'Grand Inception system'. The motives behind this movement were

[26] For supporting evidence and a general discussion of the *Zhuan Xu li*, see Chen Jiujin and Chen Meidong (1989). The statements of initial conditions given here do not occur together in any single ancient text.

[27] This is the month-count which allegedly begins with the New Year of the legendary Xia dynasty. Under the Qin and early Han the civil year began in the eleventh Xia month. From 104 BC up to the present the year has begun on the Xia New Year's day, i.e. the beginning of the first Xia month. Nowadays this usually falls some time during February.

complex, and the full story would take too long to tell here.[28] So far as they concern us directly, the facts are fairly straightforward.

In the first place, the new system required a new system origin. This was fixed at the winter solstice near the end of 105 BC. Under the new system this was stated to have occurred at midnight beginning day 1 of the sixty-day cycle, and to have coincided precisely with the conjunction which began the 11th month of the Xia count. This was also the beginning of year 51 in the sixty-year cycle, as had been the case with the *Zhuan Xu li* origin. The result of this new choice of system origin was to make all predicted instants of winter solstices three and a half hours later than under the old system, and to make all predicted mean conjunctions ten hours earlier.

There is some evidence that the original intention was to construct a quarter-remainder type system with the new origin. It seems, however, that within three or four years at most the new system was using basic constants which were not those of the quarter-remainder type to which all previous systems had belonged. The basic innovation was the length of the mean lunation, which was revised from 29 499/940 days to 29 43/81 days. The change is a very small one, since:

499/940 = 42.999/81 to three decimal places.

This new value was wholly motivated by the desire to have the numerologically significant number 81 as the divisor in the month fraction.[29] It actually makes the mean lunation length move a little further away from an accurate value.[30] Despite this change, the new system retained the equivalence:

19 solar cycles = 235 mean lunations.

This implies that the solar cycle must be 365 385/1539 days long.

These new basic constants dictated new 'resonance periods', whose lengths differed from those shared by the quarter-remainder systems. Briefly, they were:

One *zhang* 章: 19 years (the 'Rule Cycle'). As in the older systems, this period gives repetition of coincidence between a reference *qi* (winter solstice in this case) and mean conjunction.

[28] For a fuller account, see Cullen (1993).

[29] In the first place 81 is the square of the 9, the Yang number *par excellence*, and the largest number in the Chinese multiplication table: see the *Zhou bi* #A2. This was also taken to be the cubic capacity in units of *cun* of the standard pitch-pipe: see *Han shu* 21a, 975.

[30] To seven significant figures, we have:
True value: 29.53059 days
Quarter remainder: 29.53085 days
Grand inception: 29.53086 days.

One *tong* 統: 1539 years (the 'Concordance Cycle'). Solstice and conjunction coincide once more at midnight.

One *yuan* 元: 4617 years (the 'Epoch Cycle'). Solstice and conjunction coincide at midnight, and the day then beginning is once more number 1 of the sixty-day cycle.

Clearly it was only the first of these periods that found application in solar and lunar calendrical calculations during the period of less than two centuries during which the new system was in use. The basic solar and lunar constants of the Grand Inception system remained in use until the late first century AD. It was not however immune from criticism. In 78 BC, 27 years after the Grand Inception reform, the Grand Astrologer Zhang Shouwang 張壽王 submitted a memorial to the Emperor Zhao 昭 , at that time only 15 years old, in which he stated:

> The astronomical system is the great thread guiding heaven and earth, made by the Supernal Lord (Shangdi 上帝). From the beginning of the Han [up to the Great Inception reform], the dynasty made use of the Yellow Emperor's *Tiao lu li* 調呂曆 ('astronomical system harmonising the standard pitchpipes'). Now [under the new system] the Yin and Yang are no longer in harmony, and it is time to put right the faults in the astronomical system.(*Han shu* 21a, 978)

The matter was referred for further investigation to the calendrical expert Xianyu Wangren 鮮于妄人. After receiving no satisfactory answer to questions he had put to Zhang Shouwang, Xianyu Wangren set up a program of observations which lasted for three years and involved more than twenty persons. The result of this program was to vindicate the predictions of the *Tai chu li* and to show that Zhang's astronomical system was seriously in error. As the *Han shu* points out, Zhang was quite wrong to say that the early Han had used the system that he called the Yellow Emperor's *Tiao lu li,* and in any case it transpired the system to which he gave this name was in fact none other than the *Yin li* as recorded in the files of the Grand Astrologer's department.[31]

Zhang did not confine his unorthodox views to calendrical matters. He was also condemned for holding that the time of the Yellow Emperor was more than 6000 years before the time of his memorial, whereas the official view placed (presumably) the beginning of his reign 3269 years before the time Zhang wrote. He also believed that two sovereigns had held the throne who were not in the accepted sequence, one in succession to the founder of the Xia dynasty, and another 'between the Shang and

[31] If Zhang was using the *Yin li* under another name, this explains his claim that 'The *Tai chu li* is defective by three quarters of a day, and has reduced the lesser remainder [i.e. the numerator of the fractional part of the day] by 705 parts' (*Han shu* 2la, 978). The *Tai chu li* system origin was indeed three quarters of a day ahead of the coincidence of conjunction and solstice according to the *Yin li*, and this certainly involves removing 705 from the lesser remainder for the conjunction, since 705/940 = 3/4.

Zhou dynasties'. This second, he held, was a woman. Zhang seems to have been something of a crank; he was arraigned on a charge of 'Great Disrespect [to the Emperor]' equivalent in seriousness to blasphemy combined with lèse majesté, but he was eventually pardoned. Despite this good fortune he could not keep his ideas to himself and was eventually re-arrested, after which we hear no more of him.

The interest of Zhang's case is that it serves to warn us against seeing astronomical activity under the Han as existing only within the confines of that which was officially sanctioned. If a man with views like Zhang could fill the office that had once been held by Sima Qian, and from that position champion a system of the old quarter-day type against the new *Tai chu li,* what variety of thought might there not have been amongst private students of astronomy? This is a point to which we shall return when considering the origins of the *Zhou bi.*

It was about the time of the affair of Zhang Shouwang that Liu Xiang 劉向 (77 BC–6 BC) was born. He was a member of the imperial clan, and served for a short time as Superintendent of the Imperial Household. Several works by him are still extant, but the literary task for which he is best remembered is his labour in collating and editing texts of works for the imperial library. He also wrote on calendrical astronomy:

> In the time of the Emperor Cheng 成 [r. 32 BC to 7 BC] Liu Xiang gave a comprehensive account of the six systems of astronomy [presumably the six ancient quarter-day systems, see above] and set out their good and bad points in his *Wu ji lun* 五紀論 'On the succession of the Five [Powers]'. (*Han shu* 21a, 979)

Only fragments now exist of this work, and also of his *Hong fan zhuan* 洪範傳 'Account of the Great Plan [chapter of the ancient *Book of Documents*]' which also contained material on astronomy. From his son Liu Xin 劉 歆 (*c*. 50 BC–AD 23) much more survives of relevance to the present study. Liu Xin completed his father's work on the imperial library catalogue, a summary of which is preserved as chapter 30 of the *Han shu.* In scholarship he followed the group of Han scholars who were concerned with the historical significance of ancient texts understood in human terms, rather than as inerrant scripture to be ransacked word by word for esoteric meaning.

As a youth, he was a friend of Wang Mang 王莽 (45 BC–AD 23), who rose to power through his family connections with the empress. In AD 9, after having been the effective ruler for some years, Wang caused the boy who held the throne as nominal Han sovereign to go through a ceremony of abdication in his favour, and took the throne as emperor with the title Xin 新 'New' for his dynasty (*Han shu* 99a, 4099–100). Liu Xin became one of Wang's chief ministers, and provided intellectual support for Yang's claim to be patterning his government on the model of antiquity. Despite his membership of the old imperial clan Liu survived a number of political purges during the next three decades, including some involving his own children, until in AD 23 he was executed after the discovery of a plot to assassinate Wang and restore the Han to power.

As the founder of a new dynasty, it was imperative for Wang to adopt a new system of cosmic symbolism, including an astronomical system. By this time the view of the succession of the Five Powers had changed from that held under the early Han, where the powers had succeeded one another by 'conquest' in the sequence Earth, Wood, Metal, Fire, Water (see page 2). In l04 BC the founders of Han had decided that the dynasty ruled in virtue of the power of Earth, having conquered the watery power of Qin. The theory of Liu Xiang and Liu Xin, however, held that the dynastic succession of the Five Powers was not by conquest but by natural succession 'as from a mother to her child', in the order Wood, Fire, Earth, Metal, Water; in this sequence it was believed that the Han ruled in virtue of Fire (*Han shu* 25b, 1270–1). It was therefore natural to claim that since the last Han ruler had abdicated peacefully to Wang Mang, he must rule by virtue of the power which succeeds naturally to Fire, which is Earth. At his accession, therefore, Wang adopted yellow vestments and gave precedence to yellow in all ritual business; he also placed the beginning of the year in the twelfth Xia month rather than the first, where it had been since the time of Emperor Wu's Grand Inception reform (*Han shu* 99a, 4095).

Wang did not however sponsor the creation of an entirely new astronomical system. Liu Xin gave him the next best thing, which was an improved version of the Grand Inception system, preserving the basic solar and lunar constants while adding methods for predicting planetary movements. The result was the *San tong li,* 三統曆 the 'Triple Concordance system'. As with earlier and less ambitious systems, the *San tong li* began calculations from a 'system origin' at which all the elements were taken to be at initial values: this usually involved a general conjunction of all celestial bodies. As we have seen, it is pointless to ask whether this actually happened at the time of system origin, which might be many years before the system was constructed. What mattered was whether the use of the system origin together with the constants chosen for celestial periods produced acceptable predictions for astronomical events during the system's currency.

The greater the number of celestial bodies involved in the hypothesised general conjunction, the further back in time one had to look to find it. According to the account in the *Hou Han shu*:

> In the time of Wang Mang, Liu Xin made the Triple Concordance [system], which went back 31 *yuan* 元 before the Grand Inception [system origin], and obtained a conjunction of the five planets in a *gengxu* 庚戌 year, which it took as its Ultimate Origin. (*Hou Han shu, zhi* 3, 3082)

Since one *yuan* is 4617 years, it is obvious that Liu Xin's system origin took him back to a time well before the beginnings of recorded history, in either the modern or the Han dynasty understanding of sense of the term.

Liu's whole scheme was underpinned by an elaborate scheme of numerological cosmology. An account by Ban Gu 班固 based on Liu Xin's rationale of his system is preserved in *Han shu* 31a, 979ff. and 31b. We cannot tell how exactly Ban Gu follows Liu's original wording. Earlier, when Ban quotes Liu Xin's summary of the proceedings of a conference of experts on metrological numerology held under Wang Mang, we are explicitly told that Ban Gu has 'excised false wording, and selected true principles' (*Han shu* 32a, 955). But for our present purpose the ways in which Liu elaborated the Grand Inception system are not of major importance.

The return of the quarter remainder

Wang Mang died in his burning palace when his short-lived dynasty fell in AD 23. It was not until AD 25 that a member of the Han imperial family was once again more or less firmly on the throne. The new emperor is known to history by the posthumous title Guangwu 光武 'Brilliant Martial Prowess'. There would have been little spare time or energy for the establishment of a new calendrical system in the early years of his reign, and in any case there was no ideological motive for such an innovation. The Guangwu emperor was after all restoring an old dynasty, not founding a new one. The basic structure of the Triple Concordance system devised for Wang Mang was still that of the Grand Inception system produced for Emperor Wu in the greatest days of the Han, and there was therefore no over-riding reason to treat it as politically tainted.

As a system of mathematical astronomy, it was however beginning to show signs of strain. Even if the solar and lunar positions from which the Grand Inception system started in 104 BC had been precise, by more than a century later the errors in the values given to solar and lunar periods had accumulated steadily until their effects were obvious. We are not talking here about matters such as an error of a day or so in the prediction of the date of the summer solstice: even professional astronomers would have been unlikely to have been much disturbed by this alone, assuming their observations were careful enough to detect the error with confidence. It is clear from the historical record that the cause for concern stemmed from a much more obvious failure to predict lunar phases correctly.

It has already been pointed out that at this period Chinese astronomers could do no more than predict the mean lunation. But so long as this is followed fairly closely, it is unlikely that the actual appearance of the sky would reveal glaring errors. The sky would be moonless on the predicted day of conjunction, and in general for the days on either side of it. This was no longer true in the early years of the Han restoration:

> The [basic constants of] the Triple Concordance system had been in use from the first year of the Grand Inception reign period [104 BC], so it had been in action for more than a century. The calendar gradually got behind the heavens, so that the actual conjunction was occurring in advance of the calendar. Sometimes the conjunction would occur on the last day of the month, and sometimes the moon would [already have] appeared on the first day of the month. (*Hou Han shu*, *zhi* 2, 3025)

Our first record of a memorial to the emperor drawing attention to this unacceptable situation is in AD 32, but at the time it was thought that the empire was not yet in a settled enough condition for such matters to be attended to (*Hou Han shu*, *zhi* 2, 3025). Thirty years later, when the discrepancy had grown worse, concern was renewed. The official calendar predicted a lunar eclipse on the sixteenth day of the seventh month, but in fact it occurred on the day before. Since a lunar eclipse can only occur at the true moment of opposition, the mistake was undeniable. As a result, a group of scholars were given the task of predicting the instants of lunar phases using 'quarter-remainder methods' *si fen zhi shu* 四分之術. We are told that since they were unable to decide on a new system origin, this was all they could do. Presumably the implication is that they used the old month-length value of 29 499/940 days, and simply began the day-count from some recent and well established datum point such as the lunar eclipse just referred to. This *ad hoc* expedient did not therefore amount to the adoption of a new calendrical system (*Hou Han shu*, *zhi* 2, 3025).

Things continued to get worse. Under the emperor Zhang 章 in AD 85 it became known that the winter solstice position of the sun, an important reference point, was in fact 5 *du* away from where the system said it should be, and that in general the calendar was 3/4 of a day behind the heavens. The calendrical experts Bian Xin 編訢 and Li Fan 李梵 were given the task of sorting matters out. They evidently felt able to come to a more complete solution than their predecessors, for as a result of their labours the emperor issued an edict promulgating a new quarter-remainder system (*Hou Han shu*, *zhi* 2, 3026). In this he refers several times to the authority of the *chan wei* 讖緯 apocryphal books which seem to have become current from the time of Wang Mang. As we shall see, this point is of much importance in tracing the possible origins of the *Zhou bi*. Rather confusingly from our point of view, this system is usually called the *Si fen li* 四分曆 'Quarter-remainder Astronomical System', although as we have seen there were several other quarter-remainder systems before its time. From a solitary mention in *Hou Han shu, zhi* 3, 3057 it appears that its actual title may have been *Han li* 漢曆 'The Han system' – a name which, if common, would be as confusing as the alternative.

During the next year the new system was given some further fine tuning in respect of the alternation of long and short months, and thereafter it remained in force until the end of the dynasty in AD 220. In the subsequent Three Kingdoms period it was used for another sixty years in the kingdom of Shu 蜀, which claimed to be the legitimate Han successor state. But we have already told more than enough of the story for our present purpose.

Astronomy, cosmography and instrumentation

The *Zhou bi* is commonly thought of as the main text expounding the *Gai tian shuo* 蓋天說, 'Doctrine of Heaven as a chariot-cover' one of the three main cosmographic theories recognised in ancient China. I intend to question this view, not with the aim of asserting that the *Zhou bi* does not discuss cosmographic matters, but rather of asking just what sort of thing a cosmographic theory in ancient China might have been. One good way of doing this is to ask what a cosmographic theory could be expected to do, and why anybody should have found such a thing worth thinking about.

For better or worse, when such questions arise any Western scholar cannot avoid making comparisons with the ancient Mediterranean world. What, for instance, could be said about a comparison between the *Gai tian shuo* and the cosmography constructed by Ptolemy of Alexandria around AD 150? My own experience suggests that it is much more difficult than it seems to think comparatively without getting badly confused. In the first place, the project of making a comparison assumes that the comparanda are more or less the same sort of things – one does not normally compare a string quartet with a hamburger. Secondly, when as in the present case one is comparing the presumedly familiar (Ptolemy) with the unfamiliar (the *Gai tian shuo*) it is too easy to be overconfident of one's understanding of things that seem culturally close at hand. Before we begin to think about China, it is therefore worth pausing for a while to consider what sort of things pre-modern Western cosmographies were. What indeed is any cosmography for?

Looking back on the (pre-Einsteinian) history of cosmography during the last two thousand years, one's first impression is that the content I have provisionally allocated to this term is well chosen – the shape and size of the heavens and the earth, and the disposition and motions of the heavenly bodies. Moving backwards from Newton, through Kepler, through Copernicus (leaving Tycho in a perhaps undeserved backwater for simplicity) to Ptolemy, the story seems to be about spatial positions, displacements and velocities. But why should these things have seemed important to thinkers so diverse in their presuppositions? I suggest that what we have here is perhaps a case of backwards projection of our own assumptions.

In the Newtonian universe, which most of us still inhabit when we are not forced to remember relativity, dynamics rules all. The machine can start rolling as soon as we know the masses of the bodies in the universe and have full information about where objects are and what they are doing when the game starts. The obvious demand for a cosmography is therefore that it should give a full set of spatio-temporal data for dynamics to start from.

By the time we get back to Kepler we can no longer feel so confident that we know what is going on. The ellipses of the planetary orbits round the sun are still there, but the familiar dynamics has gone. Instead, the *Mysterium Cosmographicum* rationalises the relative sizes of the orbits by attempting to show that they fit between the series of elementary regular solids, and Kepler goes on in the *Harmonice Mundi* to suggest that their speeds can be related to the musical scale. Is it safe to assume that the spatio-temporal statements made by Newton and Kepler are simply different answers to the same question? With Copernicus (that is, the real Copernicus rather than the author of the imaginary simplified schemes of popular histories) my sense of alienation deepens. The sun is still in its familiar central position, but it becomes more and more difficult to feel that the cosmographer is simply trying to give us a snapshot of the disposition of the major constituents of the cosmos in time and space. One of Copernicus' constraints, that his scheme should predict the observations of a terrestrial observer, is familiar enough. But what of the other commitment which we meet for the first time in our backwards journey, that all motions shall be produced by combinations (however complex) of uniform circular motions? Faced with the amazing elaboration of the resulting celestial machinery, it becomes difficult to believe that Copernicus is really telling us that the universe is actually like that, in the literal and grammatical sense.

When we get back to Ptolemy, the complexity of epicycles, deferents and eccentrics diminishes somewhat. But in fact the problem we met with Copernicus is worsened. Ptolemy is clearly in simple earnest when he argues that the earth is a sphere, and that it is at rest in the centre of the universe.[32] But what can he mean by a lunar theory which predicts the path of the moon through the sky quite well, but makes its distance from the earth vary by a factor of two although to the naked eye its apparent size is constant?[33] Ptolemy's account is ostensibly realist, but much of what seems at first sight to be the apparatus of a physical cosmography looks more like an essentially computational structure, at least as far as the super-lunary part of the cosmos is concerned.[34]

Perhaps we need to clear our minds a little on the question of what Greek astronomers were up to before we compare them with their Chinese colleagues. We are familiar

[32] Toomer (1984), 4–7, 40–5.

[33] Neugebauer (1975), vol. 1, 88, and Lloyd (1973), 127–8.

[34] Thus in most of Ptolemy's writing 'all planetary orbits are treated separately, assuming for each one of them a radius $R = 60$ for the deferent', Neugebauer (1975), vol. 2, 917.

with the standard story of the development of Greek astronomical thinking: Eudoxus (active 365 BC) was, as Dicks says

> the first Greek astronomer of whom we have direct evidence that he worked with and fully understood the concept of the celestial sphere, and the first to have attempted the construction of a mathematically based system that would explain the apparent irregularities in the motions of the sun moon and planets as seen from the earth. (Dicks (1970), 153)

Later Apollonius (*c.* 200 BC) provided the kinematic apparatus of eccentrics, epicycles and deferents that replaced Eudoxus's homocentric spheres, and with Hipparchus (*c.* 150 BC) a flow of information from Babylonia provided a firm basis of systematic data which Ptolemy's scheme exploited fully. Now it is a commonplace that Babylonian astronomers operated without (so far as we can tell) any significant scheme of cosmography as an underpinning to their complex arithmetical computations, which related largely to horizon phenomena. It is usual to take it that their role in Ptolemy's work was simply to provide reliable numbers. But it may well be that this evaluation is due to our own presupposition of the primacy of spatial cosmography along what we take to be typically Greek lines. A Babylonian might well have seen Ptolemy's work differently, and I think that most Chinese astronomers would have tended to agree with him.

Cosmography in ancient China

Like all ancient cultures, China had a rich fund of stories to explain the origins of human civilisation and the universe within which human beings existed. For the present purpose, however, we need not concern ourselves with the Chinese equivalents of Atlas in ancient Greece or of Tiamat in Babylonia. Such material needs the methods of the mythographer rather than the historian of science.

Our concern is largely with views that were advocated on grounds that were to some extent explicitly and systematically argued. In many aspects of ancient Chinese thought the most creative period is the three centuries from 500 to 200 BC, roughly from the time of Confucius to the rise of the Han. However, so far as the surviving texts show, the shape, size, arrangement and motions of heaven, earth and the celestial bodies do not seem to have become the centre of much debate before the first century BC. But by about AD 180 Cai Yong 蔡邕 could write:

> There are three schools which talk about the body of heaven. The first is called *Zhou bi* 周髀; the second is called *Xuan ye* 宣夜; the third is called *Hun tian* 渾天. The tradition of the *Xuan ye* has been broken off. The computational methods of the *Zhou bi* are all extant, but if they are compared with the celestial phenomena

> they are mostly in error, so the astronomical officials do not use them. Only the *Hun tian* gets near the truth: the bronze instruments used by official astronomers on the observatory platform are made after this pattern ... the officials possess the instruments, but lack the original books [of the *Hun tian*]. ... I have searched for old texts for years on end without finding any. (*Hou Han shu*, *zhi* 10, 3217, comm.)

This is not the place for a full discussion of the cosmographical debate in China. But we clearly cannot avoid following up Cai Yong's mention of the term *Zhou bi*, and we shall see that the *Hun tian* exerted a significant influence on the book now called the *Zhou bi*. As for the *Xuan ye*, Cai Yong makes it clear that he claims no knowledge of it, and we may as well take him at his word. There are certainly no extant earlier accounts of this term or of any ideas said to be linked with it.[35]

Our previous discussion of pre-modern Western cosmographies has put us on guard against assuming that the 'three schools' *san jia* 三家 were simply three competing answers to a single question about the spatial layout of the universe. In listing them in the way he does, it seems likely that Cai Yong was simply following a standard Chinese doxographic procedure going back as far as the end of the second century BC. At that time Sima Tan 司馬談 arranged the documents of pre-Han thinkers into six schools, *jia*.[36] The result has been a source of confusion for many later scholars, since Sima Tan's listing has led scholars to treat (for instance) a well-defined and self-conscious group such as the Mohists on the same terms as the miscellaneous group of cosmologists, diviners and 'masters of [special] arts' *fang shi* 方士 who are usually lumped together as the 'Yinyang school'.[37]

We can hope to avoid such confusion in the present case by asking what circumstances produced the doctrines and practices referred to here as the *Zhou bi* and *Hun tian*, and what purposes they served. I shall suggest that the key to understanding their different roles and statuses lies in the relations between three ingredients of astronomical theory and practice:

(1) Cosmographic theory.
(2) Computational procedures .
(3) Observational instrumentation and procedures.

To a considerable extent these topic headings are paralleled by the way the ancient Chinese astronomical literature is structured. It is not just the content of these three

[35] There are later texts which claim to be able to give us information about the *Xuan ye*, but it is suspicious that the later the texts the more we are told.

[36] *Shi ji* 130, 3288–9.

[37] See Graham (1989), 377f.

categories, but also their relations of interdependence which seem to vary from culture to culture and through historical time. For Babylonian astronomers, so far as we can tell, computational procedures presupposed no cosmographic theory of any importance, and were directly linked with naked-eye observations of critical horizon phenomena such as the day of appearance of the first lunar crescent above the western horizon just after sunset. For Ptolemy, Copernicus and Tycho computational procedures flow directly from the structure of what are at least ostensibly accounts of the geometrical/kinetic structure of the cosmos. Instrumentation is neutral between the different approaches of these men: Ptolemy and Copernicus might have envied Tycho's lavish equipment, but they would have been agreed that the measurement of angles between the sight-lines of a terrestrial observer was common ground from which they all started.

In China the relations of the three aspects of astronomy were different. If one is forced into a crude dichotomy, one might say that Chinese computational procedures were Babylonian rather than Ptolemaic: predictions are generated by arithmetical procedures rather than by kinematic geometry. On the other hand I hope to show that the kinds of numbers generated by Chinese astronomical computation are after a certain point significantly affected by cosmographic change. The real root of the difference comes in the relation of the third division of astronomical theory and practice to the other two. It is my contention that the differences between the so-called *Hun tian* and *Zhou bi* 'schools' in ancient China can be most clearly understood in connection with changes in instrumentation and observational procedures. The reader should therefore be warned that the discussion which follows is structured with this view in mind.

The primacy of time

At the beginning of this introduction, I pointed out that the primary role of early Chinese calendrical astronomy was 'to bestow the seasons [*shi* 時 literally "times"] on the people'. In my view the early phase of Chinese astronomical activity, which extended up to around 100 BC, was primarily concerned with the heavens as the source of a series of events ordered in time rather than as a spatially integrated whole. I do not of course mean that no-one ever looked at the night sky and saw the stars in constellation patterns against which the moon and planets moved. What I do mean is that we will best understand the work of early mathematical astronomers if we think of them as primarily concerned with time rather than space.

To some extent this point hardly needs arguing at all. As we have already seen the job of early calendrical astronomers was simply to say on what days the mean lunar phases would fall, and to keep the lunar year in reasonably close step with the cycle of the seasons. It was not a basic demand of the calendar that they should say (for instance) where amongst the stars the moon would be found on the first day of the month. But as I have already mentioned there was from quite early on a stellar reference system for describing the positions of celestial bodies – the sequence of the 28 'lodges' *xiu* 宿. How can such a system be thought of as primarily a basis for temporal rather than spatial ordering? To understand this point we need to consider the actual observation procedures within which the system functioned.

Lodges, gnomon and waterclock

Histories of Chinese astronomy describe the system of the *ershiba xiu* 二十八宿 '28 lodges' in two ways, each of which is in my view open to the same basic criticism. Sometimes we are told that the lodges constitute an equatorial division of the heavens (Needham (1959), 231). In that case it becomes hard to see why the determinative stars of the lodges bear so little clear relation to the position of the celestial equator at any plausible epoch. A more cautious but still misleading type of account limits itself to saying that the system divides the sky into unequal slices of right ascension, rather like the segments of an orange (Cullen (1980), 43). While these descriptions are adequate for most of the history of pre-modern Chinese astronomy, they lead to much confusion when we come to consider the origins of the system and its early use.

In both cases the fundamental flaw in these accounts is that they speak as if it can be taken for granted that early Chinese astronomy used the concept of the celestial sphere, with its equator and hour angle circles. But it is quite possible to do a good deal of astronomy without any such conception at all. So far as our evidence goes, Babylonian astronomers managed perfectly well without it (Dicks (1970), 169–70).[38] As Dicks points out, the innovation of using the celestial sphere as a framework for astronomical calculation and observation is a decisive event in the history of Greek astronomy, datable to some time before the work of Eudoxus *c.* 365 BC, and nothing but confusion can result from assuming its presence in the mental furniture of earlier astronomical thinkers (Dicks (1970), 161).

What evidence do we have of when the concept of the celestial sphere entered the minds of Chinese astronomers? We have a good guide to the thinking of an official astronomer around 100 BC in the *Tian guan shu* 天官書 'Monograph on the Celestial Offices' in Sima Qian's 司馬遷 universal history *Shi ji* 史記 '*Records of the Historian*'. Like his father Sima Tan 司馬談, Sima Qian held the office of Tai shi 太史 'Grand

[38] Neugebauer (1975) vol. 1, 348 speaks of 'the working of a theoretical astronomy which operates without any model of a spherical universe, without circular motions and all the other concepts which seem "a priori" necessary for the investigation of celestial phenomena'. On the other hand vol. 2, 577, veers between caution and deducing a spherical heaven from the use of 'orthogonal coordinates for bodies close to the ecliptic' – but vol. 1, 547 rejects any notion of spherical astronomy decisively, at least so far as the Normal Star texts are concerned. Even for Neugebauer's 'System A' it is significant that although measurements which would be the equivalent of latitude are discussed they are not given in degrees, but in a different unit, the 'barleycorn' which is apparently a measure of length: vol. 1, 514.

Astrologer', the principal astronomical post at the Han court. Throughout this detailed discussion of the layout of the sky and the movement of the celestial bodies we find no mention at all of the terms *chi dao* 赤道 'red road' [=celestial equator] or *huang dao* 黃道 'yellow road' [=ecliptic] so familiar from later writings as the most important great circles of the celestial sphere.[39] We are not, however, dealing with astronomy in a period before quantitative measurement: thus for instance in a discussion of the motions of Jupiter we are told:

> When Jupiter first appears, it moves eastwards 12 *du*, coming to a halt after 100 days. Then it turns retrograde, and moves 8 *du* backwards. 100 days, and it moves eastwards again, moving 30 *du* and 7/16 *du* in a year. (*Shi ji* 27, 1313)

The path of the planet is not named in any way. It is significant that the angular measure *du* 度 is only used for east–west displacement; when north–south displacement or displacement above the horizon is in question, linear measures such as *chi* 尺 'foot' (*Shi ji* 27, 1331) or *zhang* 丈 'ten feet' (*Shi ji* 27, 1334) are used. We seem to be in a situation rather like that of the Babylonians mentioned elsewhere. In this case, fortunately, we know rather more about the astronomers in question than we do about their colleagues in Babylonia. We do in fact have a description of the activities of a major project involving Sima Qian himself, the great *Tai chu* 太初 calendrical reform of 104 BC. When they started work, Sima Qian and his colleagues

> Fixed east and west, erected gnomons and set waterclocks working, so as to determine the separations of the 28 lodges in the four quarters [of the heavens] (*Han shu* 21a, 975)

This description neatly encapsulates an impression that we could reinforce from many other sources, including the *Zhou bi* itself: the primary instrument of the early Chinese astronomer was the gnomon, whether used to measure solar shadows, or as here for stellar observations. To understand the role of the gnomon in relation to the stars, we have to recall that unlike the Babylonian astronomer who was primarily interested in horizon phenomena, his Chinese counterpart took special note of what we would call meridian transits. Like the archetypal Chinese emperor on his throne, the Chinese astronomer 'faced south' and observed which heavenly bodies crossed his

39 The nearest we get to anything that could be a reference to the ecliptic as the average path of the planets through the sky is the statement that 'When the moon follows the *zhong dao* 中道"middle road", there will be peace and harmony' (*Shi ji* 27, 1331). But the Tang commentator takes this as no more than a reference to the path taken by the moon through the asterism Fang (π Scorpii and other stars). A similar reference occurs at *Shi ji* 27, 1299, where the reference is clearly to an individual asterism rather than to a complete track round the heavens.

north–south sightline as the diurnal rotation of the heavens carried them from east to west.[40] A body which is just on the meridian is said to be 'centred' *zhong* 中.

As already mentioned, the most ancient complete repertoire of 'centred stars' *zhong xing* 中星 is to be found in the *Yue ling* 月令 chapters of the third century BC *Lü shi chun qiu* 呂氏春秋. For each month we are given the name of one lodge which is 'centred' at dusk when the stars first become visible, and one which is 'centred' at dawn just before they disappear in the light of the rising sun. Clearly in such a case we are only dealing with fairly rough indications, particularly since the position of a particular lunar month relative to the cycle of the seasons (and hence to the cycle of dusk or dawn transit stars) can 'float' by up to thirty days even if intercalation is being run efficiently. For such purposes I doubt whether a formal sightline would have been essential, but a gnomon and a well-defined north–south line as described in *Zhou bi* section #G would clearly help to reduce ambiguity.

But how does one determine the 'separations' *xiang ju* 相距 of the 28 lodges quantitatively, and how does one define the position of other bodies within the lodge system? The answer clearly lies in the waterclock used by Sima Qian and his colleagues. The use of this instrument for timing star transits is confirmed in another account of Han practice:

> The orifice vessel serves as a clepsydra, and the floating arrow serves [to mark] the *ke* 刻 [there were 100 *ke* to the day]. Set the clepsydra running and count the *ke* to observe the centred stars （*Hou Han shu*, *zhi* 3, 3056)

About 140 BC Hipparchus seems to have been working on generally similar lines. The final chapter of his *Commentary to Aratus* lists stars culminating as closely as possible to one hour intervals; this has been taken to suggest that Greek astronomers could time such an interval by waterclock to within a half minute of accuracy.[41] By simply timing the intervals between the transits of the determinative stars of lodges, it is easy to determine their width in *du*, given that the total width of all lodges must be 365 1/4 *du*. The longest interval that would have needed to be timed was about two hours in the case of the lodge Well, which extends over some 30 *du*. Some smaller lodges would pass the meridian in about 15 minutes. The figure of 365 1/4 *du* is another case of a time interval appearing to us as angle – it is plainly derived from the number of days required for the sun to return to its starting place among the lodges,

[40] The connection between the southwards position of the ruler and the observation of meridian transits of stars is made explicitly by Liu Xiang 劉向 (77 BC–AD 6) in his *Shuo yuan* 說苑, 18, 3a in the *Sibu congkan* edn.

[41] See Dreyer (1953), 161, and Neugebauer (1975) vol. 1, 279. For an account of some surviving Han dynasty clepsydras see Chen Meidong (1989). It is of course unlikely that we will ever recover an example of a clepsydra used by early Chinese astronomers; those we do find will most probably be those used for bureaucratic or domestic purposes.

precession being of course ignored at this period. In terms of spherical astronomy, this timing process is equivalent to determining the differences of the determinative stars in right ascension.

The point, however is precisely that we must not think in terms of spherical astronomy, since it looks very much as though early Han astronomers and their predecessors had not yet begun to think of the heavens as spherical. Those measurements in *du* which we unconsciously translate into angular terms seem for them to have been concerned with time intervals rather than spatial measurements. A measurement in *du* essentially tells you the time interval between the meridian transits of two celestial bodies on the same night. Indeed, if it was not for the fact that the times of dusk and dawn change from day to day, a measurement of (say) 10 *du* could be interpreted very simply as meaning that ten days would elapse between the dusk or dawn meridian transits of the bodies concerned. It is a plausible conjecture that such approximate measurements actually lie behind the origin of the lodge system and the use of the *du* as a celestial measure.

Given only the evidence so far discussed, it is already fairly clear that it is not safe to assume that early Han astronomers such as Sima Qian thought in terms of the celestial sphere. But apart from what we can learn from the *Zhou bi* itself, we also have concrete evidence in the form of what are clearly schematic models of a cosmography which correlates with gnomon and waterclock transit observations. These objects are the subject of the next section.

The shi *cosmic model*

Xi 犧 and He 和, the archetypal official astronomers, were (it will be recalled) charged with delivering 'the seasons [*shi* 時, literally "times"] to the people'. Throughout the history of traditional China, the problem of selecting the most favourable time for a given activity was also the principal task of diviners, often operating at less exalted levels of society than court astronomers. From at least the early Han dynasty such diviners used instruments known as *shi* 式. Specimens of these objects excavated in recent decades reveal a pattern of thought that is clearly related to the astronomical practices outlined above, as well as to the contents of the *Zhou bi*.[42]

To some it may seem odd that equipment used for what we would now call fortune-telling might have any relevance to the history of cosmographic thinking, or to the history of science generally. The most obvious objection to such an attitude is that

[42] I have discussed my views on these objects at length in Cullen (1981); detailed arguments will not be repeated here. For wider background, see also Loewe (1979), chapter 3. My article in *Early China* followed Harper (1979). Connoisseurs of academic polemics may enjoy the resultant exchange in *Early China* 6 and 7. Others may wonder why the authors found it worth the energy required. Li Ling (1991) is the first part of a continuing study, and gives a very full review of the literature and archaeological evidence, although it omits Kalinowski (1983).

it is based on anachronistic assumptions about the attitude of pre-modern societies towards divination. No detailed argument on this point is necessary here. In the ancient Mediterranean world one has only to think of the Stoics, for whom a faith in the possibility of divination was a necessary consequence of their conviction that the cosmos was a unified whole in which an unbreakable chain of cause and effect could be discerned. For ancient China, as we have seen, the status of the Book of Change from the Han dynasty onwards shows us how far divinatory schemes were bound up with the cosmology that expressed the ideology of the centralised imperial state.[43]

More importantly, however, a dismissive attitude towards divination ignores the possibility that technical 'specialists' *fang shi* 方士 such as diviners may have been just the people who were likely to have concerned themselves with the topics on which our interest now centres. Angus Graham has argued this point in connection with the origins of Yinyang and Five Phase concepts:

> Cosmological speculation, which is at the beginning of Greek philosophy, entered the main current of Chinese thought only at the very end of the classical period. Down to about 250 BC it belongs to a world right outside the philosophical schools, that of the court historiographers, astronomers, diviners, physicians, and musicmasters [. . .]. (Graham (1989), 325)

The strong dissent registered by Nathan Sivin (1992) must be noted, and in any case cosmology is not cosmography. But if the cosmographical discussions we begin to discern from the first century BC onwards have roots in earlier thought, these roots are certainly not located in any of the mainstream pre-Han philosophical works which have come down to us.[44] We should not therefore be surprised when the material evidence points in a different direction, and this is what I suggest the *shi* does.

Figure 2 shows a specimen excavated from a tomb dated about 165 BC. The object is in two parts, the round heaven-disc pivoted at its centre, and the square earth-plate beneath it. On the heaven-disc are marked the names of the 28 lodges in sequence at equal intervals, with a sequence of twelve numbers written against some of them. These numbers indicate the lodge in which the sun should be found for the month in question. The centre of the disc bears a conventionalised representation of the stars of the Northern Dipper, α to η Ursae Majoris. The names of the lodges are repeated round the outer band of the earth-plate, seven to each side of the square. Thc middle band bears the twelvefold set of cyclical signs, the *di zhi* 地支 'earthly stems' while the innermost band carries the ten *tian gan* 天干 'heavenly branches' in modified sequence to fit the twelve spaces available.

[43] Graham (1989), 358ff.

[44] However, in *Later Mohist Logic, Ethics and Science*, 23 and 369–71 Graham speculates that the Mohists may have produced works on geometrised astronomy, which were either eventually lost, or perhaps in part at least eventually incorporated into the *Zhou bi.*

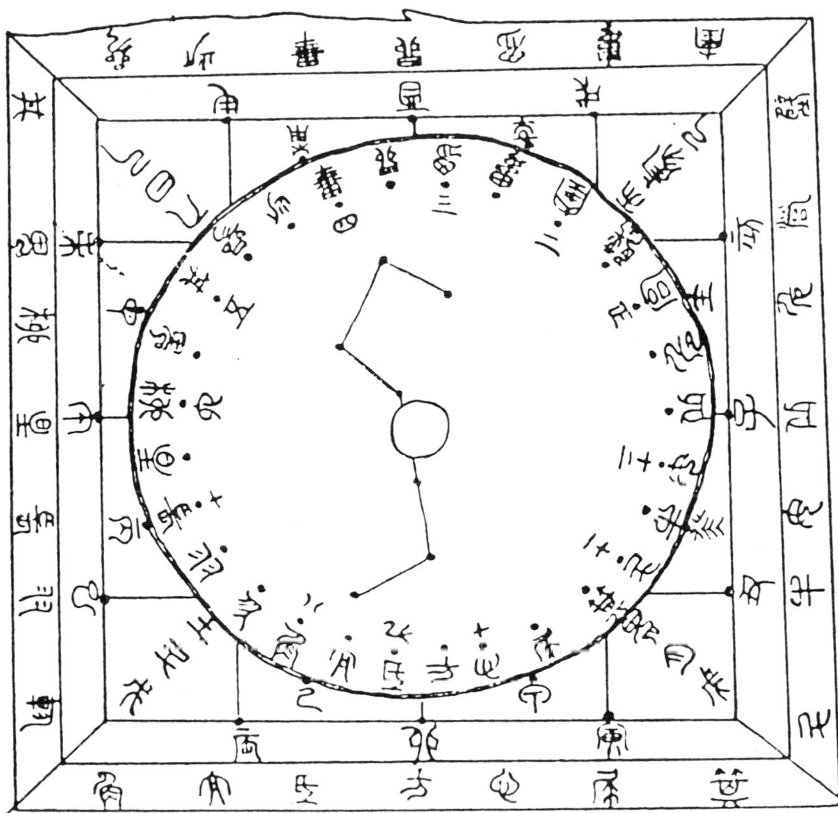

Figure 2. *Shi* cosmic model, early second century BC, from *Kaogu* (1978), 5, 340; line drawing of original object. The actual width of the earth-plate is 13.5 cm.

What we have here is clearly a model of heaven and earth in some sense. But what sense precisely? There are certainly some spatial elements present. The parts of the model representing heaven and earth are shaped to fit a common early Chinese convention repeated in paragraph #A6 of the *Zhou bi*:[45]

> The square pertains to Earth, and the circle pertains to Heaven. Heaven is a circle and Earth is a square ...

[45] See also, for instance, *Huai nan zi* 淮南子 1, 4b, also 7, 2a and 15, 3a (*Sibu congkan* edn) around 120 BC. A section of the *Da Dai liji* 大戴禮記, compiled from earlier material around AD 100, makes a point of denying that heaven and earth could really be different shapes as the common saying implies, which shows that the idea of a round heaven and square earth must have been widespread: see the *Hanwei congshu* edn 5, 7b.

The *shi* is (to modern eyes at least) a highly conventionalised representation of the ancient Chinese cosmos. The most obvious example of conventionalisation is the way in which the lodges are shown on the heaven-disc as having equal extents: in reality their widths should vary from 5 *du* for the lodge Ghost to 26 *du* for Well. In another example of conventional representation the constellation of the Northern Dipper *bei dou* 北斗 (the Great Bear) is shown at the centre of the disc, with the pivot (which presumably represents the north celestial pole) close to the stars ε and δ Ursae Majoris. In reality, around 200 BC these stars were over twenty degrees from the pole – about forty moon-widths. The point of this arrangement for the diviner who used this device was probably that it symbolised the great importance of the Dipper as the wielder of cosmic power. As Sima Qian himself put it:

> The Dipper is the Imperial chariot; it rotates in the centre, and looks down on and regulates the four regions. The division of yin and yang, the establishment of the four seasons, the balancing of the five phases, the shifting of the nodes and degrees, and the fixing of all sequences – all are bound up with the Dipper. (*Shi ji* 27, 1291)

No-one could argue that Sima Qian was unaware of the plain fact that the celestial pole does not lie within the Dipper. Indeed he opens his description of the sky with the clear statement that the constellation known as *Bei ji* 北極 'north pole' contains the bright star that we now know as β Ursae Minoris, which in his time was within a few degrees of the pole (*Shi ji* 27, 1289). If even he could be moved to speak of the Dipper as 'in the centre', it is no surprise to find a diviner actually putting it in that position quite literally.

What we have here is clearly not primarily intended as a planisphere designed to show us the actual appearance of the heavens. Its purpose is simply to give us a schematised model of the basic time sequences of the cosmos. The diviner is not concerned with the angular distances between asterisms, but merely with their sequence, and with the order in which the sun moves through them. Knowing this information, the device can be set at an appropriate orientation for any instant, and the state of the cosmos can thus be assessed.

Let us take the position shown in figure 3 as an example. We are some way through the sixth month, and so we place the sun between the lodges numbered six and seven. That puts it at the right hand end of the horizontal diameter of the disc. It is thus opposite the cyclical sign *wu* 午 on the middle band of the earth-plate, corresponding to the direction due south. The time of day shown thus corresponds to noon, when the sun makes its meridian transit – in Chinese terms, it is 'centred'. As well as representing

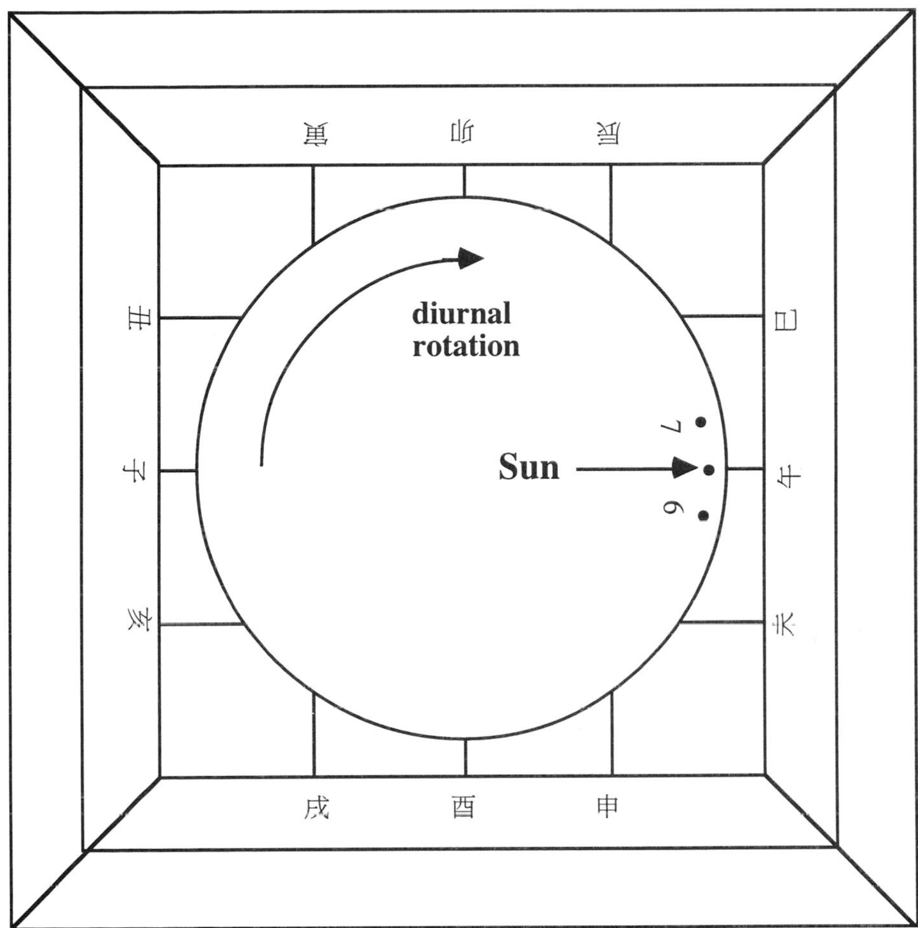

Figure 3. The *shi*, showing situation in the sixth month: for clarity only essential graduations are shown.

the direction due south, *wu* also represents the 'double-hour' *shi* 時 from 11 a.m. to 1 p.m. As time passes we would have to rotate the disc clockwise to keep it in correspondence with the heavens, until at midnight the position of the sun was opposite the sign *zi* 子 in the middle of the left hand side of the earth-plate, corresponding to the direction due north, and to the double-hour 11 p.m. to 1 a.m. As the disc rotates, a sequence of lodges will pass by the position of the sign *wu*, and thus will make their meridian transits in turn. Although this device clearly does not attempt to represent astronomical data quantitatively, the sort of phenomena it represents are precisely those with which Sima Qian and his colleagues concerned themselves, as we have seen.

But what has this to do with the *Zhou bi*? In the first place, it turns out that it is possible to establish a very direct connection indeed. At one point in section #G the text describes a procedure which involves marking out a large circle on level ground, followed by the graduation of its circumference into 365 1/4 *du,* each *du* taking up one *chi* of circumference. A gnomon is erected at its centre, and with the aid of a further (movable) gnomon at the circumference a series of sightings is taken on the standard stars of the 28 lodges, with the object of determining the width of each. The method proposed involves (in our terms) a crude approximation of differences in right ascension to changes in azimuth. It could not have yielded useful results, and the motives for including it in the text were other than practical ones.[46] A section of the text reads:

> #G7 [60b]
> When [the lodge] Well is centred at midnight [on the winter solstice], the beginning of Ox falls over the middle of *zi* 子 [due north]. When Well is 30 *du* and 7/16 *du* to the west of the central standard gnomon, and falls over the middle of *wei* 未 [30 degrees w. of south], then the beginning of Ox falls over the middle of *chou* 丑 [30 degrees e. of north]. With this, heaven and earth are matched together.

These statements bear no more than a rough resemblance to reality in the sky of the first century BC. The text assumes that the standard stars marking the initial points of the lodges Well (μ Geminorum) and Ox (β Capricorni) are (in modern terms) diametrically opposite in right ascension, so that when the first is crossing the meridian due south of the observer the other is due north of him, although out of sight over the northern horizon. In the first century BC the right ascensions of these stars were close to 65 degrees and 276 degrees respectively, so that the alignment claimed was missed by over thirty degrees, the equivalent of two hours of time.

On the heaven-disc of the *shi* shown in figure 4 however, the lodges are laid out at equal intervals round the circumference and the alignment is very close to that specified by the *Zhou bi*. Thus, when the dot marking the position of Well on the disc falls over the cyclical sign *wu* on the right side of the earth-plate, which corresponds to the south, the dot marking Ox is close to the sign *zi* on the left hand (north) side of the disc. When Well has shifted through 30 7/16 *du* (i.e. 30 degrees) to align with *wei*, Ox

46 As pointed out below, the idea seems to have been to suggest that *gai tian* methods could do everything the *hun tian* could do.

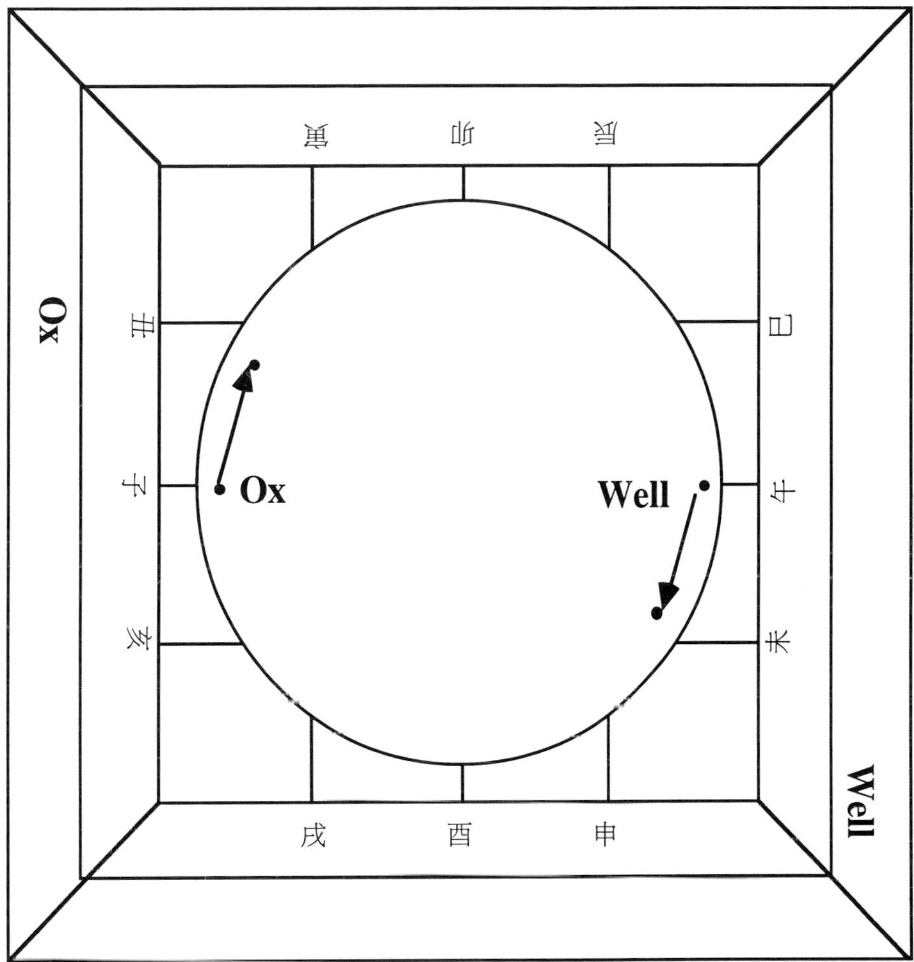

Figure 4. Alignments as in *Zhou bi.*

has shifted through an equal angle and aligns with *chou.* All this corresponds precisely to what section #G of the *Zhou bi* tells us.

Most significant is the statement that when this has occurred heaven and earth are 'matched' *xie* 協. It is precisely under these circumstances that the 28 lodges marked on the heaven-disc are in proper correspondence with their names round the edge of the earth-plate. Such a striking coincidence suggests strongly that the thinking of the author of this section of the *Zhou bi* was conditioned by the schematic view of the universe embodied in the *shi* cosmic model. When we turn to the physical model of the universe discussed in the *Zhou bi* we shall find stronger connections still.

The gai tian *cosmography*

We now turn to the details of what is usually called the *gai tian* 蓋天 theory of cosmography; the term *zhou bi* 周髀 is also used. The significance of the first term is clear enough: *gai* means the large umbrella-like canopy over an ancient Chinese chariot, and heaven was thought to be related to earth in the same way that this canopy was related to the body of the chariot. Thus in a fragment of a poem by Song Yu 宋玉 written *c*. 300 BC we read:

> The square earth is a chariot (*che* 車);[47]
> The round heaven is its canopy (*gai* 蓋).
> (*Beitang shuchao* (repr. Taipei 1962) 149.3b)

More graphic still is the description in the *Kao gong ji* 考工記 'Artificers' record', a pre-Qin document which now forms part of the *Zhou li* 周禮 'Rites of Zhou':[48]

> The squareness of the chariot's body images the earth; the roundness of the canopy *gai* images heaven; the thirty spokes of the wheels image days and months; the 28 spokes of the canopy image the stars [of the 28 lodges] (*Zhou li zhu shu* 40, 7a–7b)

And the text continues to describe the chariot's banners imaging the stars of the four quarters. Clearly the ruler who travelled in such a cosmic vehicle was expressing a claim to more than mere earthly power. More trivially, the analogy between the lodges and the spokes of an umbrella is obviously to be preferred to the (to us) more natural analogy of orange segments (see above) with its misleading suggestion of a celestial sphere.

I think it is clear that the term *gai tian* 'canopy heaven' has links with the ancient image of the cosmos seen in the texts quoted here. So can we say that the *gai tian* theory goes back to the time of the Warring States? We need to be cautious at this point. In the context of the early history of science, the word 'theory' brings with it many associations that may prove confusing and anachronistic. Do we, for instance, want to talk about Tycho's cosmography and the ancient Egyptian picture of the

[47] The notion of the earth as a vehicle also seems to lie behind the expression *kanyu jia* 堪輿家 'those who specialise in the [load] bearing carriage' used to refer to experts in siting – the discipline popularly known as *fengshui*. 風水.

[48] This book, also known as the *Zhou guan* 周官 'Offices of Zhou' was traditionally thought to be a description of the state organisation of the early Western Zhou *c*. 1000 BC. The offices of state in this work are correlated with the four seasons. By the second century BC the last section, the 'winter office' had been lost and was replaced by the *Kao gong ji* 考工記. Jiang Yong 江永 (1681–1742), a teacher of Dai Zhen 戴震, argued that the language of the work showed that it could not be of Western Zhou date, and was probably written in the state of Qi. See the preface to his *Zhou li yi yi ju yao* 周禮疑義舉要 'Selected points on problems of meaning in the *Zhou li*', vol. 101 p. 765, *Siku quanshu* photographic reprint.

sky-goddess Nut arching herself over the earth as both being 'theories' in the same sense? It seems to me that the description 'cosmographic theory' is best reserved for use when a view about the structure of the universe is being put forward in a context of debate or polemic (whether actual or potential) in which one has to envisage the replies of opponents who may ask for clarification of one's views, attack them directly, put forward competing views of their own, or perhaps even seek to change the ground on which the debate is taking place.[49] Now in my view some of the ideas about heaven and earth which are eventually identified as constituting what we may call the *gai tian* theory go back to a time before there was any controversy on such questions. Under such circumstances people do not attach labels to their views: they simply say 'heaven and earth are like this ...'.

I shall argue that the self-conscious advocacy of one cosmographic view against another is an intellectual phenomenon that probably did not appear in China until the first century BC. On that basis, Song Yu's brief reference should not be taken as showing that he was a partisan of the *gai tian* theory: he was just saying what everybody else knew about heaven and earth, as is the *Kao gong ji*. The same must apply to the statement we find in the *Lü shi chun qiu* 呂氏春秋, a compendium of learning assembled in the state of Qin in 239 BC:

> The pole star moves together with heaven but the pivot of heaven does not move.[50] At the winter solstice the sun moves along the most distant track. It moves round the four poles, and its decree is called dark and light. At the summer solstice the sun moves along the closest track and reaches the highest point. Beneath the pivot there is [then] no [alternation of] day and night. (*Lü shi chun qiu* 13, 3b, *Sibu congkan* edn)

In the light of what we know about early Chinese astronomical practice, and the *shi* cosmic models which were based on it, we can already make good sense of this passage without having to turn to the *Zhou bi* at all. Song Yu's poem (about 60 years older than the work we are quoting) and the *Kao gong ji* both give us further help,

[49] On this view, there is no difficulty in talking (say) of Aristotle putting forward the theory that the earth is a sphere at rest at the centre of the universe (*De Caelo* II.xiii). Aristotle is so keenly aware of the possibility of alternative views that he reviews and attempts to refute them with great care. A modern scientist is not likely to find himself in agreement with Aristotle's criteria for deciding which view is correct, but the point is that like a modern scientist Aristotle has such criteria and applies them explicitly.

[50] This is a true statement, since the pole star of any epoch is in general likely to be a perceptible distance away from the pole itself. The *Zhou bi*, section #F, actually sets out a method by which the movement of the pole star may be observed. The word used for this movement is *you* 游 as in the text translated here. However the commentary on this section of the *Lü shi chun qiu* by the third century commentator Gao You might be interpreted to imply that his text referred to the pole star as stationary. The point is not important for our present purpose.

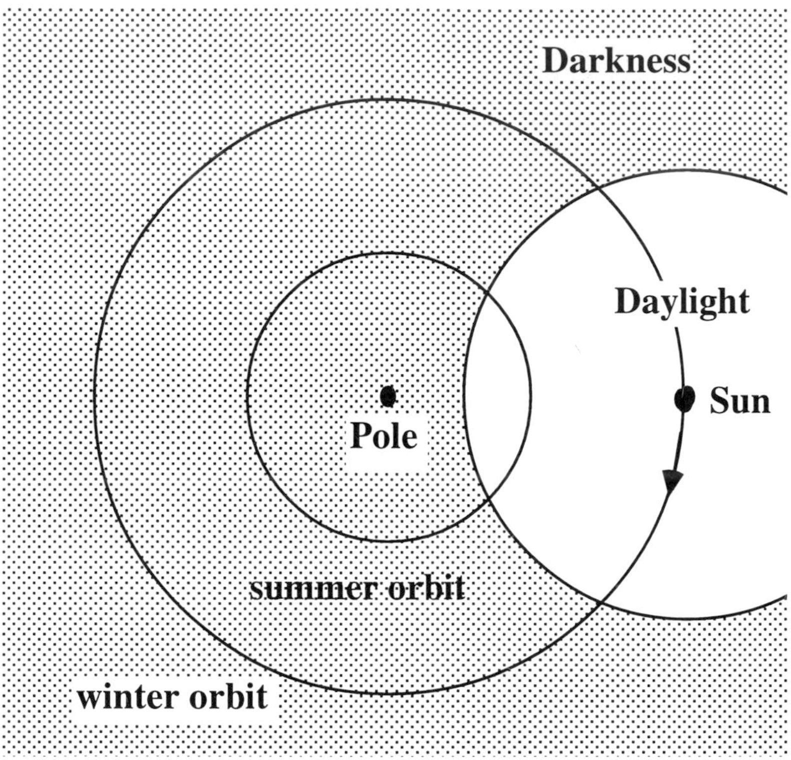

Figure 5. Cosmography of *Lü shi chun qiu* .

with their vivid picture of the great round umbrella canopy mounted on its vertical pole above the square body of the chariot. Like the canopy, heaven is pivoted about its centre, near which is the pole star. It rotates once daily above the stationary earth, carrying the pole star, the sun and by implication the other heavenly bodies with it. When the rotation of heaven carries the sun over a particular region of the earth beneath, it is daytime there. When it moves away from over that region night falls. The seasons are neatly explained by the notion that in winter the sun is furthest from the axis of heaven, and hence will be further away from the more central regions of the earth than it is in summer, when it is close to the axis. The cosmography presupposed in the few sentences translated here relates closely in plan view to the appearance of the *shi* models: see figure 5. In elevation we obviously have to take it that heaven is some distance above the earth, although the matter is not raised in the *Lü shi chun qiu.*

It is striking to find this text of the late Warring States period telling us that in summer there is no proper day and night in the regions below the celestial pole, evidently because the sun is then so close to the pole that it is always daylight there.

The same point is made in the *Zhou bi*, paragraph #B22. Although the modern astronomical explanation would be different, the fact remains that at the Earth's north pole there is indeed perpetual daylight from the spring to the autumn equinoxes. At present it is impossible to tell whether these statements are the consequence of travellers' tales from the far north, or whether they are simply the consequence of a predictive success by a view of the cosmos conceived on the basis of purely Chinese experience.

Let us sum up the situation so far. I have argued that Chinese astronomers of the second century BC, as yet without the concept of the celestial sphere and following the paradigm of meridian transit observation, naturally saw themselves as primarily involved in measurements of time intervals rather than of spatial intervals on the heavens. I have suggested that the *shi* may be seen as a physical expression of this approach to the cosmos. I have also suggested that some early thinkers were actually prepared to say that heaven and earth were physically related to one another in the same way as the heaven-disc and earth-plate of the diviner's instrument. As we shall see, the views they held were those which were later to be known as the *gai tian* theory. But almost as soon as we are able to form a picture of the complex of observational practice and schematic thought involved, a disturbing influence comes on the scene and an age of polemics opens.

The hun tian *and the mapping of space*

The revolution in practice

There was a significant alteration in the character of Chinese astronomy between Sima Qian's work around 100 BC and Cai Yong's account of the three cosmographic theories around 180 AD. This change is associated with the adoption of what Cai refers to as 'the *hun tian* 渾天', and it may be seen as a threefold process:

(1) Astronomers came to perceive the heavens as a vast rotating sphere, with themselves at the centre, rather than as a rotating umbrella-like cover overhead, with themselves some distance away from its axis.

(2) New observational instruments were adopted, consisting of combinations of graduated rings with adjustable sights, making up what are now called armillary spheres.

(3) In observational technique, the emphasis shifted from timing successive meridian transits to the measurement of angular separations between heavenly bodies, conceived as distances between points on the celestial sphere. Instead of being a specialised measure of what were essentially time intervals between transits, interpreted as increments of rotation of the heavenly disc, the du was generalised

so that it could be used (for example) for the angular distance from a given star to the north celestial pole.

The first two innovations were inextricably bound up together as complementary aspects of the change in astronomical practice outlined in (3). We may recall that the situation in the ancient Mediterranean was rather different. Despite the deep changes in astronomical theory between Eudoxus (365 BC) and Ptolemy (AD 150), all Greek astronomers during this period were agreed that their observational data fundamentally consisted of angles between the sightlines of a terrestrial observer. Ptolemy had better instruments for making such observations than Eudoxus (who some scholars feel had none at all), but the differences in instrumentation between the two astronomers had no essential connection with their differences in astronomical theory.

In my view, however, the neat distinction between cosmographical theory and instrumentation does not hold in China at this period.[51] There is therefore little point in trying, as some scholars have done, to find ways of telling whether an otherwise unspecified reference to 'the *hun tian*' points to *hun tian* instruments, or to the cosmographical theory so named. The presence of one entails the presence of the other, since they were both essential parts of a fundamentally new approach to astronomical practice involving the measurement of space rather than time.

This discussion is not a general history of the development of early Chinese astronomical theory and practice, but must necessarily limit itself to the minimum necessary to make sense of the development of the *Zhou bi*. Nevertheless, to do this properly we need to follow the story of the coming of the *hun tian* in its essential outlines. To see this story most clearly I suggest that instead of looking first for occurrences of the words '*hun tian*' we should look first for signs of the changes in astronomical practice that I contend were the more fundamental aspects of the overall process. As in the case of the *gai tian*, it is by no means certain that the existence of a particular way of doing astronomy does not considerably predate the coining of a descriptive label for it.

What, then, is the kind of thing we shall be looking for? In the first place, as I have already mentioned, an infallible sign of the presence of the spherical heaven in the astronomer's mind is when he begins to use the *du* 度 as a generalised angular (or at least pseudo-angular) measure for the separation of points on the heavens. Secondly, the emergence of the celestial equator and ecliptic as great circles of equal status is something that only makes sense in the context of a spherical heaven. Thirdly, and by no means least, we expect to find references to the development of instruments designed to measure angles in the new way.

[51] Thus when in section #G the *Zhou bi* attempts to find ways to smuggle measurements of north polar distance in *du* into the *gai tian* cosmos it can only do so at the cost of considerable subterfuge and distortion, since its cosmography rules out the use of *hun tian* instruments.

We shall shortly see that the first century BC was a critical period for such innovations as these. There are however few if any signs of this in the astrological and calendrical monographs of the *Han shu* 漢書, compiled by Ban Gu 班固 (AD 32–92) and his sister Ban Zhao 班昭 (AD ?48–?116) as a history of the Western Han dynasty (202 BC–AD 9). One reason for this may be that Ban Gu based his account on government records that did not fully reflect innovation at the margins of official astronomical activity. Much of his material on mathematical astronomy and associated topics is, as he states (*Han shu* 21a, 955 and 979) simply taken directly from the official reports of Liu Xin 劉歆 (*c.* 50 BC–AD 23). Nothing in these sections shows signs of the new approach.

The situation is rather different in the *tian wen* 天文 monograph of the *Han shu*, which appears to have been compiled by Ban Gu himself, using Sima Qian's monograph in the *Shi ji* as a basis for the listing of constellations. It is here that we meet the first reference in the standard histories to what seems to be the ecliptic. But despite the appearance of this term, there is considerable doubt as to its precise significance in Ban Gu's thinking. The relevant passage deserves quotation at length:

> The sun has its Middle Road; the moon has its Nine Tracks.[52] The Middle Road referred to here is the Yellow Road. One [source] calls it the Brilliant Road. In the north [quarter of the heavens] the Brilliant Road reaches [the lodge] Eastern Well, when it is closest to the north pole; in the south [quarter] it reaches [the lodge] Ox, when it is furthest from the north pole; in the east [quarter] it reaches [the lodge] Horn, and in the west [quarter] it reaches [the lodge] Harvester, and [at these two times] it is midway [between its maximum and minimum distances] from the pole. At the summer solstice it reaches Well, and is in the north close to the pole, so the shadow is short. If one sets up an eight *chi* gnomon, the shadow length is one *chi*

[52] The 'nine tracks' *jiu xing* of the moon are further explained by Ban Gu after his discussion of solar motion: *Han shu* 21a, 1295. It is there made clear that the nine tracks follow a straightforward scheme of correlation of seasons (each divided into two), directions and colours. Eight of them may be explained as follows:

Season:	Spr. I/II	Sum. I/II	Aut. I/II	Wint. I/II
Quarter of heavens:	East	South	West	North
Colour:	Green	Red	White	Black

The ninth track, corresponding to the centre (the Middle Way, i.e. the ecliptic) correlates with yellow and does not fit into the seasonal scheme. Such arrangements were a commonplace of Han thought: see Graham *Disputers of the Tao*, 351, 382. Any attempt to read into this a reference to such matters as the 18-year cycle of the lunar nodes, or the nine-year cycle of the lunar line of apsides would obviously be a gross anachronism. All that can be deduced from Ban Gu's discussion is that he is aware that the moon does not always follow the same track as the sun. For an indication of the inchoate state of Chinese ideas on lunar motion around the time that Ban Gu wrote, see the memorial by Jia Kui 賈逵 *c.* 96 AD, *Hou Han shu, zhi* 2, 3029–30, and elsewhere in the same chapter.

> five *cun* and eight *fen* [= 1.58 *chi*]. At the winter solstice it reaches Ox, and is far from the pole, so the shadow is long. If one sets up an eight *chi* gnomon, the shadow length is one *zhang* three *chi* one *cun* and four *fen* [= 13.14 *chi*].[53] At the spring and autumn equinoxes the sun reaches Harvester and Horn. It is midway [between its maximum and minimum distances] from the pole, so the shadow is midway; if one sets up an eight *chi* gnomon the shadow length is seven *chi* three *cun* and six *fen* [= 7.36 *chi*].[54] These are the differences in the sun's distance from the pole, and the rules for the length of the shadow. Distances from the pole are hard to ascertain, so one has to use the shadow. It is the shadow that enables one to know the sun's north and south [displacements]. (*Han shu* 21a, 1294)

It is certainly a little hard to know what to make of this. The path of the sun through the constellations is certainly given a name – three names in fact – and the second of these names is the one which in later accounts definitely refers to the great circle of the ecliptic on the celestial sphere. There must however be some doubt as to whether that is how Ban Gu thinks of it. He specifically rules out the possibility of measuring the sun's north polar distance, which is nowhere given an value in angular measure, and states instead that one must rely on gnomon measurements. This is precisely the approach of the *Zhou bi*, which also states that the sun is in the lodges Ox and Well at the solstices, and that it is then at its extreme distances from the pole.[55] A further passage may be quoted to reinforce the impression that Ban Gu is still doing astronomy the old way:

> The motion of the sun cannot be pointed out and known [i.e. by direct observation]. Therefore one uses the [dusk centred] stars at the two solstices and two equinoxes as means of observation ... [there then follows a listing of the degrees of each lodge which should cross the meridian at dusk at the relevant times of the year, with divinatory interpretations of the significance of these phenomena occurring early or late]. (*Han shu* 21a, 1295)

Clearly however, anybody with an armillary instrument could in fact observe the position of the sun directly, either by measuring its north polar distance as it crosses

[53] Taking the solstitial shadows together, they imply a latitude of observation of 34.9 degrees north, and a value of 23.7 degrees for the obliquity of the ecliptic. Both are reasonable values for the region of the Han capitals and the first century AD.

[54] The equinoctial shadow length given here is clearly not an observed value, but is simply the mean of the solstitial values. The same procedure is followed in the *Zhou bi*: see for example section #H, which has the value 7.55 *chi* as the mean of solstitial values of 13.5 and 1.6 *chi*.

[55] Paragraph #D5. Section #G is of course trying to do more than Ban Gu when it produces its faked method for measuring north polar distances of the sun at the solstices and equinoxes, quantities which are not mentioned at all in the *Han shu*.

the meridian, or (in conjunction with a waterclock) deducing in which lodge it was located.

Taking all this evidence together, Ban Gu's only advance on Sima Qian seems to be that he gives the path of the sun through the constellations an unambiguous name. We have already seen that Sima Qian may have done that in any case, and the mere naming of the solar path is not sufficient for us to deduce the use of *hun tian* methods.[56] If Ban Gu's monograph was the only piece of astronomical writing from the first century AD, we would have to conclude that there was as yet no sign of *hun tian* astronomy. But this is certainly not the case, as we shall see.

It will be recalled that the year AD 85 marked the final abandonment of the *San Tong Li* 三統曆, and the reintroduction of a calendrical system of the quarter-remainder type. A few years later in AD 92, the year of Ban Gu's death, we have a memorial by Jia Kui 賈逵 on matters relating to innovations in calendrical astronomy, which is preserved for us in the monograph by Sima Biao 司馬彪 (third century AD) now to be found in the *Hou Han shu*. It does not take very long before we recognise that Jia Kui's material shows unambiguous evidence of *hun tian* practices. Near the beginning of his memorial he is discussing the position of the sun at the winter solstice:

> Mr. Shi's 石 [i.e. Shi Shen's 石申] *Xing jing* 星經 'Canon of the Stars' says: 'The Yellow Road curves through the beginning of Ox, then straightens out through the twentieth *du* of Dipper, which is 25 *du* further from the pole than the [corresponding point] on the Red Road, i.e. the twenty-first *du* of Dipper.'[57] (*Hou Han shu*, *zhi* 2, 3027)

We hardly need further evidence: Jia Kui is obviously familiar with the ecliptic, the equator (which is what 'Red Road' now clearly means), and the concept of switching reference from one great circle to the other. We also have the *du* now used to measure north polar distances. It is hard indeed to reconcile such writing with that of Ban Gu a few years earlier. One explanation that suggests itself is that, as a historian, Ban Gu based his account on what he found in Western Han sources rather than the practice of

[56] It is also notable that Ban Gu says nothing that could be taken as a reference to the celestial equator. If he had had the celestial sphere in mind, he would surely have wanted to reserve the term 'Middle Road' for that great circle rather than for the ecliptic. Indeed he actually uses the later term for the equator 'the Red Road' *chi dao* 赤道 as a name for one of the moon's 'nine tracks'.

[57] I punctuate this text with Yabuuchi (1969), 52, whereas the editors of the Beijing edition end the quote after '25 *du*'. This means that one would have to say that the twenty-first *du* of Dipper was 25 *du* from the pole, which would be grossly in error. This quantity is obviously the obliquity of the ecliptic, as confirmed by Jia Kui's repetition of this datum on p. 3029, columns 7 and 8. My interpretation of *zhi* 直 as 'straighten out' reads the text as saying that as the ecliptic passes through the real solstitial position (which for Shi Shen is clearly the twentieth *du* of Dipper on the ecliptic) it is aligned parallel to the equator: see the identical usage by Jia Kui (same reference) and in Zhang Heng's *Hun yi* quoted in *Han shu*, *zhi* 3, 3076–7 ninth column. Near the end of the essay Zhang makes the same point as Shi Shen.

astronomers in his own day a century further on. We have already seen evidence that he did just that. But, as we shall see, the tone of Jia Kui's writing suggests that towards the end of the first century AD court astronomers were regarded as less than progressive, and it may well be that Ban Gu's account was simply based on obsolete information obtained through 'the proper channels'.

We are fortunate in that Jia Kui's discussion goes on to review the history of various astronomical problems, and in so doing gives us essential information about changes in astronomical practice related to *hun tian* thinking. Thus he tells us a little later on:

> Your servant has previously submitted a memorial pointing out that when Fu An 傅安and his colleagues used the tYellow Road to measure the [positions of] sun and moon at half and full moons, they were mostly correct. But the astronomical officials, who only used the Red Road, were not in agreement with the sun and moon. They were often more than a day wrong in comparison with the present system, and all they could do was to memorialise this as being a portent, even to the extent of making out that the sun had moved retrograde! On the Yellow Road the degrees of motion turn out naturally, and no such 'portents' occur. ... I have previously answered that at the winter solstice the sun is 115 *du* from the pole, that at the summer solstice it is 67 *du* from the pole, and that at the spring and autumn equinoxes it is 91 *du* from the pole. (*Hou Han shu*, *zhi* 2, 3029)

He then proceeds to tell us more about the differences between the motions of the sun and moon when referred to the ecliptic and equator, ending by noting that all was just as Fu An had said, and that twelve experts whom he had consulted on the question concurred in saying:

> The [official] star maps are made by the 'compass method' *gui fa* 規法,[58] but the sun and moon really follow the ecliptic. The officials do not have the [appropriate] instruments, and do not know how to go about the business. (*Hou Han shu*, *zhi* 2, 3029)

Jia Kui's energetic polemic paid off, for nine years later in AD 103 the order was given to construct 'the Grand Astrologer's ecliptic bronze instrument', and we are given the extensions allocated to the 28 lodges on the ecliptic of the new device. But it seems that the astronomical officials did not welcome devices imposed on them by outsiders, for we read that the officials found the new instrument 'difficult for making observations' and largely neglected to use it (*Hou Han shu*, *zhi* 2, 3029–30).

At this point we may pause for a moment and sum up. Writing near the end of the

[58] We are perhaps to imagine some kind of planisphere, with the celestial pole at the centre and a circle centred on it graduated in *du*. Such graduations would be the equivalent of our right ascension, which Jia Kui sees as graduations marked on the equator.

first century AD Ban Gu gives an account of astronomy that is in no way inconsistent with the *gai tian* style of thought and practice followed by Sima Qian two hundred years before. The conservatism of the official astronomers of this period encountered by Jia Kui reinforces our suspicion that they might have been Ban Gu's source of information.[59] At the same time, it is clear that an essentially unofficial group had committed themselves to the *hun tian* way of doing astronomy, and were pressing for its official adoption. Having set up this relatively firm benchmark based on the evidence of astronomical practice, we can turn to the doxographical evidence with less risk of confusion.

Polemics on the hun tian

It takes two sides to have a satisfactory argument. The problem with the material just discussed is that the officials with their old fashioned views present a sitting target, whether to the modern reader's condescension or the eager advocacy of Jia Kui. Did nobody of any intellectual reputation set themselves up to defend the old views? The answer seems to be 'Yes, but not for long'. It seems that around the time of Wang Mang 王莽 (r. AD 9–23) two friends talked about these matters in some detail. Enough fragmentary references to their discussions have come down to us for us to see how one of them changed his mind. The account comes from a fragment of the lost book *Xin Lun* 新論 New Discussions' by Huan Tan 桓譚 (40 BC–AD 30):

> That most perceptive of men Yang Ziyun 楊子雲 [i.e. Yang Xiong 楊雄, 53 BC–AD 18] based himself on the explanations of the multitude of [previous] scholars, and took it that the heavens were a cover (*gai* 蓋) always rotating leftwards [i.e. when one faces the pole], with the sun moon and stars following it from east to west. So he drew a diagram of its form with the graduations of its movements, and checked it against the four seasons, the calendrical constants, dusk and dawn and day and night. His wish was to give men of his own day a standard, and to pass on a method to those who came later.
>
> I raised the following objections: At the spring and autumn [equinoxes] day and night should be equal. The sun rises due east at *mao* 卯 and sets due west at

59 Jia Kui's references to official astronomers using the Red Road (equator) to reckon the motions of the sun and moon when they should have used the Yellow Road (ecliptic) may leave us with the impression that they were still basically using a *hun tian* approach, although their instruments lacked the ecliptic circle. The implication would then be that they traced the motions of the sun and moon on the equatorial ring. This is however not likely to have been the case. As Ban Gu's account makes clear, the officials were perfectly well aware that the path of the sun and moon varied in its distance from the pole. I suggest that he is simply translating their essentially *gai tian* approach into his own *hun tian* terms. He would see their use of the old system of the *du* as a measure of the rotation of heaven as essentially the same as laying out the lodges along the equatorial ring. It does not follow that the official astronomers thought that way too.

> *you* 酉. But if we look at this from [your] cosmic perspective [we are talking about] the east–west line of the human [observer], not the east–west line of heaven. This goes through the Dipper pole, which is the pivot or axis of heaven, just like a [chariot] cover having a hub. Even though the cover can turn, its hub does not move. Likewise heaven rotates, but the Dipper pole does not move, which is how we know it is the centre of heaven. Now if we look up we see it in the north, and not directly overhead. But [that means that] at the spring and autumn equinoxes the sun rises and sets to the south of the Dipper [pole]. [Since heaven] is turning like a cover, that means that the northern [part of the sun's] track is distant from us, and the southern part is close [i.e. our east–west line is not a true diameter dividing the circle of the sun's daily motion in half]. So how could the lengths of day and night be equal? Ziyun had no explanation.
>
> Another time I had a memorial to submit with Ziyun, and we were seated under the eaves of the verandah of the White Tiger Hall.[60] Because of the cold we turned our backs to the sun, which warmed them for a while, and then the sun's rays moved away, and our backs were no longer warmed by them. I used this as an illustration for Ziyun, saying 'If heaven really turned like a cover, carrying the sun towards the west, its rays should still be shining under this verandah, but just have moved a little towards the east [rather than the sun having increased its altitude so that we are now in the shadow of the roof]. As this is not the case, on the contrary [the facts] correspond to the methods of the *hun tian* school. Ziyun thereupon destroyed [the device] he had made. {So the scholars who say that heaven turns to the left are right.}[61] (*Taiping yulan* 2, 6b–7a.)[62]

As this dialogue makes plain, the *gai tian* was doomed to failure when it was forced to step outside its original limited role and compete with a cosmography in which the prime aim was spatial verisimilitude.

This example is, so far as we know, the first and only time in the Han dynasty that a figure of wide intellectual repute comes out explicitly in favour of *gai tian* methods. Later in the first century the lonely figure of the great sceptic Wang Chong 王充 is however to be found writing polemics on the losing side in his 'Discourses weighed in

[60] It will help us to understand the references to the movement of the sun if we recall that court business normally began at dawn.

[61] I suspect that this last sentence is an intruded comment by a copyist who did not understand the argument.

[62] Other material is known which bears some relation to their dialogue. See *Jin shu* 11, 282; *Sui shu* 19, 506–7. In this latter, Yang Xiong is shown (presumably after his conversion) as making eight objections against the *gai tian*; these are mostly quite cogent on physical grounds, and include such points as that from the top of a high mountain a water level reveals that the sightline to the rising sun is a little below the horizontal. Whether Yang himself actually drew up this list is something we cannot be entirely sure about.

the balance' *Lun heng* 論衡 (*c.* AD 83), but his low status and general unfashionableness in his day make his testimony of rather marginal value.[63] Nevertheless he has clearly heard of the *hun tian*, since he devotes some effort to arguing that the explanation of the sun's setting cannot really be that heaven has carried it below the earth, since this would mean that it would have to pass through the subterrene waters (*Sibu congkan* edn 11, 8a–8b). In reality heaven is flat and level, just like earth (11, 8b). The apparent rising and setting of the sun for a particular observer is an optical illusion caused by its moving closer to him and further away as heaven rotates with the sun attached to its underside: when it seems to set for one observer another perceives it as culminating. The apparent meeting of heaven and earth at the horizon is similarly a purely optical effect (11, 8b–10a). As we shall see, such views are precisely those found in the *Zhou bi* itself.

Returning to Yang Xiong, it is clear that the *hun tian* was already making itself felt by the beginning of the first century AD. Yang Xiong himself is the source of the only explicit statement taking it back any further. In his *Fa yan* 法言, a work modelled on the Confucian *Analects*, he recounts the following dialogue between himself and a questioner:

> Someone asked about the *hun tian* 渾天. He replied 'Luoxia Hong 落下閎 (fl. 110 BC) thought it out (*ying zhi* 營之[64]); Xianyu Wangren 鮮于忘人 (fl. 80 BC) gave it a scale (*du zhi* 度之); Geng [Shouchang] 耿壽昌 the palace assistant (fl. 50 BC) made a representation of it (*xiang zhi* 象之). How exact it is! No-one can contradict it.' They asked about the *gai tian* 蓋天. He said: 'The *gai*! The *gai*! It leads to difficulties and is inaccurate'. (*Fa yan* 10, 1b, *Sibu beiyao* edn)

In the light of Yang's previous discomfiture at the hands of his friend Huan Tan, one may sense a certain bitterness in his final exclamation. Working our way back in time through the people mentioned in this list, we find unsurprisingly that independent corroboration is harder to find the further back we go. Geng Shouchang is mentioned in Jia Kui's memorial of AD 92, in the context of his argument that the motions of the sun and moon could only be properly understood with reference to the ecliptic:

> In the second year of the Ganlu 甘露 reign period [52 BC], the Supervisor of Agriculture Geng Shouchang memorialised that he had measured the motions of

[63] Indeed there is some doubt how far he really knew much of what was going on, since he fails to mention the terms *gai tian* or *hun tian*, let alone *zhou bi*. He does however talk about heaven being compared to a cover, *gai*: *Lun Heng* 11, 8b, *Sibu congkan* edn.

[64] *Ying* can equally well refer to an act of mental creativity or to the construction of some physical device. In view of my doubts about the advisability of trying to separate *hun tian* ideas from *hun tian* instruments it would be pointless to argue at length as to what meaning Yang Xiong intended here – *pace* Needham (1959), 354–5 and Cullen (1980), 36 and note 39.

> the sun and moon with a 'diagram instrument' *tu yi* 圖儀 in order to examine and test the phenomena of celestial rotation. [He found that] when the sun and moon passed through [the solstitial lodges] Ox and Well, the moon moved fifteen *du* for each *du* moved by the sun, [whereas] when they reached [the equinoctial lodges] Harvester and Horn the moon moved thirteen *du* for each *du* moved by the sun. It was [measurement with reference to] the Red Road that brought this about – this fact is [thus] something that was well known well before our own time.[65] (*Hou Han shu*, *zhi* 2, 3029)

What was a *tu yi*? The term is not a common one like the standard *hun tian yi* 渾天儀 later used for armillary instruments. There are however two parallels to help us. In a memorial of AD 143 we are told that under the Emperor Zhang (apparently at the time of the calendar reform of AD 85):

> The degrees [specified in] the calendar were investigated and checked: the *tu yi*, the [gnomon] shadow and the waterclock were [all] in accordance with heaven. (*Hou Han shu*, *zhi* 2, 3037)

The fact that the *tu yi* is mentioned together with gnomons and waterclocks makes it look very much like some kind of observational instrument, and when in AD 175 Cai Yong refers to observations conducted in his own time with a *hun tian tu yi* 渾天圖儀 (*Hou Han shu*, *zhi* 2, 3039) the evidence is hard to withstand. The device used by Geng Shouchang in 52 BC was almost certainly a simple *hun tian* device of some kind. It is interesting to note that in the West Ptolemy was, like Geng, moved to construct an armillary instrument *c*. 150 AD by his need to follow the movements of the moon more exactly.[66]

But what kind of device? For Jia Kui, the whole point of mentioning Geng at all is that his instrument could only measure in the plane of the equator. An extremely simple device – and one that might appropriately be called a 'diagram instrument' – would suffice for this. One simply draws a graduated circle on paper, and fixes it to a flat board. The board is then arranged in the plane of the celestial equator. No more than the simplest sighting device would be required, such as a fixed pin perpendicular to the board through its centre, and an arrangement for a similar back-sight to be moved round the graduated circumference, perhaps by insertion into a series of holes.

All this is mere speculation, although there may be some grounds for taking it a

[65] Of course the true story of the variability of lunar motions is more complex than a simple periodicity linked to fixed longitudes, as indeed Jia Kui notes a little later (p. 3030). But he is here engaged in polemic rather than cool weighing of the facts.

[66] *Almagest* V.1, H351; Toomer (1984), 216. Ptolemy's instrument was however a fully developed device with ecliptic rings, unlike Geng's simple equatorial scale two centuries earlier. We cannot tell whether Ptolemy was the first to use such an instrument in the West.

little more seriously. The resulting object would be very much like the traditional Chinese equatorial sundial, and certain stone dials said to be of Han date do in fact carry markings and holes quite similar to those hypothesised here.[67] But speculation aside, the irreducible minimum seems to be that around 52 BC Geng Shouchang had a device that enabled him to measure angles in the plane of the equator. As I have pointed out elsewhere, Geng did not face any great practical barrier in creating such a device, since we know of a specimen of a disc set out accurately with the extent of the twenty-eight lodges in *du* from as early as the second century BC.[68] All he had to do was mount such an object in the equatorial plane and add sights. For comparison, on the other side of the world Hipparchus (*c*. 135 BC) mentions the existence of a fixed equatorial measuring-ring of bronze in the Square Stoa in Alexandria.[69] Of course, once anybody had set up such a ring, it would be highly suggestive in cosmographical terms: no-one making use of it could resist the implication that half of the heavens were out of sight beneath the earth at any instant. Once again we see the essential unity of *hun tian* theory with observational practice.

Of the other two men in Yang Xiong's list there is less to say. Xianyu Wangren is recorded as leading the defence of the Grand Inception system against an attack made on it in 78 BC.[70] He headed a team of twenty men who carried out a program of observation over two years. There is nothing to indicate what sort of instruments they may have used, but he was certainly in a good position to gather data for 'giving a scale' to a new measuring device, whatever may have been meant by this expression.

The only reliable data on Luoxia Hong are those which link him with the Grand Inception reform, and as we have seen the only pieces of observational equipment mentioned in connection with that particular project are gnomons and waterclocks.[71] Sima Qian, who was in a position to know, tells us that:

> the *fang shi* 方士 Tang Du 唐都 divided the sectors of heaven *fen qi tian bu* 分 其 天部, while Luoxia Hong 落下閎 from Ba 巴 prefecture [in Sichuan 四川] carried out calculations to revise the calendar *yun suan zhuan li* 運算轉暦 (*Shi ji* 26, 1260)

This gives us no reason to accept or reject Yang Xiong's claim that Luoxia Hong was connected with the early history of the *hun tian*.[72] In general it is hard to know

67 Needham (1959), 302–9, particularly figs. 128, 129, 131 and 132.

68 Discussed in detail in Cullen (1981), 34f.

69 Quoted by Ptolemy (*c*. AD 150) in the *Almagest* III.1, H195; Toomer (1984), 133. Ptolemy himself refers shortly afterwards to 'the [two] bronze rings in our Palaestra, which are supposed to be fixed in the plane of the equator' (III.1, H197; Toomer (1984), 134).

70 *Han shu* 21a, 978.

71 See *Shi ji* 25, 1260 and *Han shu* 21a, 975.

72 While there are other texts that purport to link Luoxia Hong with the construction or use of armillary

quite what value to set on his whole story, which paradoxically might have seemed more plausible if the figures he had named had been unknown to the historical record. As it is, one has the vague suspicion that he has simply picked three eminent astronomers of the preceding century and unloaded responsibility onto them. And to make matter worse, there is the further suspicion that Yang's wish to imitate passages from the *Analects* means that he has to have several names on his list, with different roles ascribed to each of them.[73]

So how far back can we go in our attempt to trace the origins of *hun tian* astronomy? Whatever our doubts about Luoxia Hong around 100 BC, Geng Shouchang seems a reliable benchmark around 50 BC. We may add to this the indirect evidence of Yabuuchi Kiyoshi's analysis of data in the *Xing jing* 星經 'Star Canon', a fragment of which was quoted by Jia Kui in part of his memorial quoted above. On the basis of this analysis it seems highly likely that around 70 BC someone in China had an instrument at his disposal enabling him to determine north polar distances – a distinctively *hun tian* feature.[74] It seems therefore that we can be reasonably confident of finding the *hun tian* well back into the first century BC.

At the other end of our time scale, we have seen that by the time of Jia Kui near the end of the first century AD effective propaganda was being made to force the use of *hun tian* instrumentation and practice upon an evidently reluctant or simply incompetent astronomical establishment. Shortly after Jia Kui's work, near the beginning of the second century, we have two important *hun tian* writings by the polymath Zhang Heng 張衡, in which the fully developed form of the cosmography is set out, together with a description of a *hun tian* armillary instrument. A full discussion of these documents would take more space than is justified in the present context, which is after all confined to outlining the historical setting for the *Zhou bi*. Long sections have been translated by Joseph Needham, to whom the interested reader is referred.[75] A sketch of Zhang Heng's conception of the universe is shown in figure 6.

The *hun tian* universe as shown here was in essence a sort of giant planetarium designed to reproduce the phenomena seen by an observer in the latitude of the

instruments, I have argued elsewhere that they are corrupt or too late to be treated as independent evidence. The corrupt text is a story (in itself slightly dubious) about Yang Xiong talking to an old workman who had made an armillary instrument: see *Taiping yulan* 太平御覽 2, 11a. In the confused version of *Beitang shuchao* 北堂書鈔 130, 12a the old workman becomes Luoxia Hong himself, a chronological and social impossibility. The other source is a third century AD work quoted in the Tang *So yin* 索隱 commentary to *Shi ji* 26, 1261, which says that Luoxia Hong *yu di zhong zhuan hun tian* 於地中轉渾天 'rotated a *hun tian* [instrument] at the centre of the earth [i.e. the Chinese capital]'. This is clearly inspired by a combination of Yang Xiong's statement that Luoxia Hong was the inventor of the *hun tian* with the words of Sima Qian quoted above. See Cullen (1981), note 46.

73 See *Analects* XIV.9.

74 See Yabuuchi (1969), 46–95.

75 Needham (1959), 216–7, 355–7.

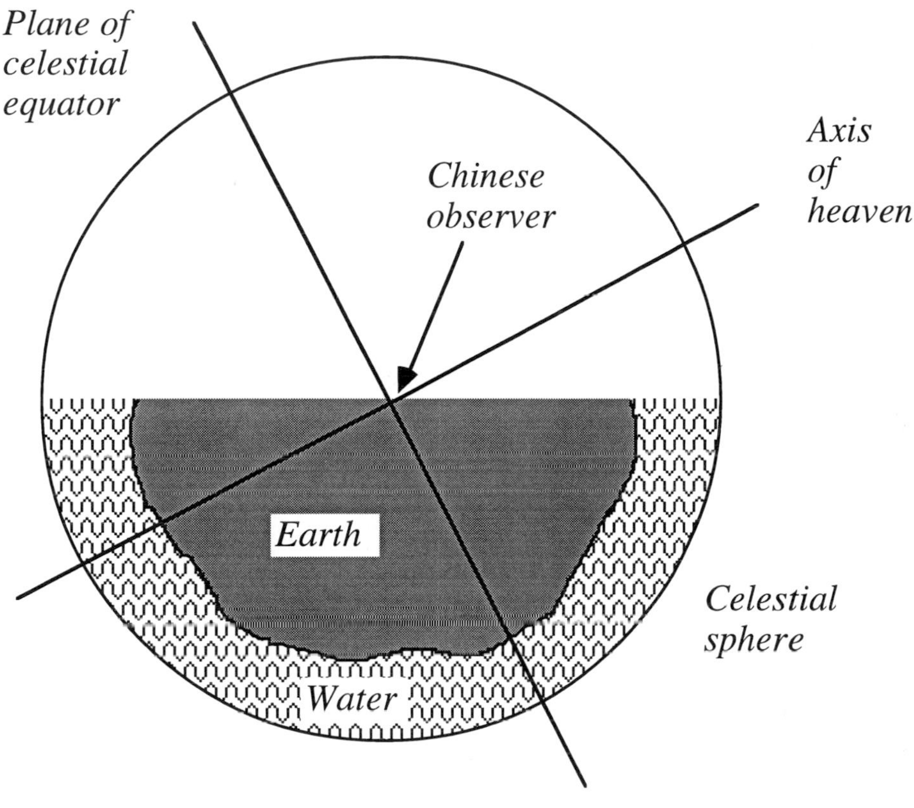

Figure 6. Zhang Heng's *hun tian* universe.

Yellow River basin. It is a 'one-sphere' universe, in which everything depends on the daily rotation of the celestial sphere about its inclined axis, carrying the celestial bodies up over the eastern edge of the flat earth, and then down again over the western edge. It is notable that once this scheme had been described in general terms neither Zhang Heng or any later writer felt it necessary to explain any mechanism for the detailed motions of the celestial bodies on the sphere which carried them. It is impossibleto tell how Zhang Heng derived his figure of over two million *li* for the diameter of the celestial sphere: this is roughly equivalent to one hundred thousand kilometres. For comparison, the actual diameter of the earth is only 12 800 km. While we may initially be tempted to smile at Zhang's ethnocentric assumption that China was at the centre of the universe, everybody else in the inhabited world as he knew it was for all practical purposes also at the centre.

Retrospect

We have now sketched in the background necessary for understanding the historical context of the *Zhou bi*. In this process, it has become clear that there were a number of significant changes in astronomical theory and practice in the period from the beginning of the Han up to the first century AD. Separating them under two provisional headings, we may list them as follows:

(1) In calendrical astronomy, the abandonment of the old quarter-remainder system type in 104 BC, and its replacement by a new system with its basic divisor of 81. In AD 85 this system was itself replaced by a new quarter-remainder system in the context of a lively debate in which calendrical matters, scriptural arguments and politics were all involved.

(2) In observational techniques, the appearance from sometime in the first century BC of new instruments and methods involving the conception of the celestial sphere. It does not however seem to have been until the end of the first century AD that these new ideas were forced upon the consciousness of astronomical officials, who till then remained wedded to the old techniques using gnomons and waterclocks, and, it is simplest to assume, to the conception of the cosmos that went with them.

We will now turn to a detailed examination of the contents of the *Zhou bi*, after which it should be possible to form a judgment on the origins and significance of the work in its proper historical context.

2

The *Zhou bi* and its contents

The text

On the whole the *Zhou bi* is not textually problematic, at least not in the sense that a document such as the Mohist Canon *Mo jing* 墨經 is problematic. There are certainly several instances where the text is obviously in disorder or otherwise corrupted, but so far as we can tell these do not affect crucial issues in the interpretation of the work as a whole.

The story of how the various current versions of the *Zhou bi* came down to us will be told in chapter 3. As explained there, the best available critical text is that prepared by Qian Baocong 錢寶琮 for his edition of the Ten Mathematical Classics *Suan jing shi shu* 算經十書. For the present, my purpose is simply to give the reader a brief outline of the way the text is structured, and how the accompanying apparatus is related to it.

The divisions of the text

Paragraphs and sections

In my translation the text is divided on two levels. I must bear editorial responsibility for imposing both these structures on the material. In the case of the smaller-scale division, the paragraph, I have tried to find easily manageable units of discourse at a level somewhat above the sentence. In the form in which I have created them, the paragraphs are usually terminated where the first commentator, Zhao Shuang, interjects some explanatory text. On the other hand I often group together a series of short passages that seem to me to belong in a continuous unit, despite the fact that Zhao separates them by a few characters of commentary – we must remember that the invaluable device of the footnote for commenting without interruption was unavailable to him. This division into paragraphs is simply a device to help the reader find his or her way through the text and facilitate reference, rather than an attempt to restore the structure of some hypothetical original text. The only exceptions to this are a number of instances where the text is clearly disordered enough for it to fall apart into disconnected fragments without any need for an editor to exercise any very subtle discernment.

Above the level of the paragraph, I suggest that the text divides fairly naturally into eleven main sections. Apart from the fact that the traditional division of the present text into two chapters coincides with the boundary between the fourth and fifth of my sections, there is usually little explicit indication of such divisions in the original. I have argued the case for each section division in the introduction to the translation of that section, and these arguments are assembled in summary form on page 140ff. In general my decision is based on some abrupt change of topic or style, in some cases coinciding with the point at which a pre-announced list of questions has been answered. The issue of how far these *ad hoc* divisions may actually correspond to documents with separate origins will be discussed later.

To make it easier for the reader to find the original of my translated text in Qian's edition, I have added page and column references after my paragraph numberings. Thus, the paragraph designated #D6 [47d], informs the reader that it is the sixth paragraph of section #D, and that it commences on page 47 of Qian's edition of the *Zhou bi*, in column (d) of that page, i.e. the fourth column counted from the right.

One chapter or two?

As already mentioned, current texts of the *Zhou bi* divide it into two chapters. However, so far as I can tell there is no evidence that this division was a feature of the original Han text seen by Zhao Shuang, and I do not therefore believe there is any reason to try to analyse the content and origin of the *Zhou bi* on this basis: see page 164.

Text and commentaries

The text of the *Zhou bi* is accompanied by three main commentaries, datable to the third, the sixth and the seventh centuries AD. In traditional Chinese fashion, passages of commentary are inserted into breaks in the text, which is written or printed in vertical columns. One text section could thus in principle be followed by three portions of commentary, and since each commentator evidently knew of his predecessors' work some of this material may be commentary on commentary. In general the situation is less complex than this.

The three commentaries do not, on the whole, add a great deal to our understanding of the main text in its context of origin. Most of the time the meaning is clear enough on its own, and where the text is obscure I cannot think of many instances where a commentator succeeds in making a really helpful suggestion. The main interest of the commentaries lies in what they reveal of the commentators' attitudes to the text, and of their own understanding of it.

Zhao Shuang

As will appear below (page 161), there is reason to suppose that Zhao Shuang 趙爽 may have written in the third century AD in the southern state of Wu 吳, one of the 'Three Kingdoms' into which China was divided at the end of the Han. Wu lasted from AD 222 to 280. Most of Zhao's commentary consists of the usual mixture of paraphrases of the original wording, glosses of more difficult words and background information that a Chinese teacher would traditionally give his students when reading a text with them. As a sample of the former, at the start of section #E (beginning of the second chapter according to the traditional division) where the text reads:

> #E1 [53b]
> The rotation of the sun and moon around the way of the four poles: *fan ri yue yun xing si ji zhi dao* 凡日月運行四極之道
>
> Zhao comments:
>
> *Yun* 運 'move/revolve' here means the same as *zhou* 周 'go round'. *Ji* 極 'pole' here means the same as *zhi* 至 'limit'. The reference is to the outermost *heng* 衡 [i.e. the daily path of the sun at the winter solstice]. The sun and moon go round the four regions. When they reach the outermost *heng* they turn back. That is why [the text] says 'pole'.

Apart from this, Zhao also often gives details of the process of calculation leading to results stated in the main text without explanation. At its most complex, such working might consist of instructions to square two numbers, add the results and take the square root. Zhao assumes that his readers need no help in performing the arithmetical operations he specifies.

In addition to his running commentary, Zhao makes certain substantial additions to the text. These essays have been translated and discussed in appendices 1 to 4. As he tells us in his preface, he also added diagrams to illustrate points in the text, *yi jing wei tu* 依經為圖. These diagrams found in the Southern Song reprint of 1214 are as follows:

(1) Three diagrams illustrating applications of Pythagoras' theorem, inserted between paragraphs #A3 and #A4 of the main text. These are followed by an essay in which Zhao discusses the mathematics behind the diagrams. Comparison of these illustrations with Zhao's texts makes it plain that the second and third of these diagrams do not fit

Zhao's text at all well. They may well have been added by Zhen Luan. Qian Baocong attempts to reconstruct what he believes to have been the original illustrations.[76]

(2) A diagram following paragraph #B11, entitled *Ri gao tu* 日高圖 'Diagram [for finding] the height of the sun', followed by a few sentences of explanation by Zhao. This material seems to have no obvious relation to #B11 itself, but is fairly loosely linked by its content to #B10, in which mention is made of the differences in noon shadow lengths between gnomons lying on the same north–south line. Once again a corrupt diagram has been restored by Qian.

(3) Two diagrams, one showing a square inscribed in a circle and the other showing a circle inscribed in a square. These occur between #C1 and #C2, and Zhao does not seem to have added any further explanation, if indeed he ever saw them. As noted in the main translation, the captions on the diagrams seem to be reversed.

(4) There is a further diagram, with the title *Qi heng tu* 七衡圖 'Diagram of the seven *heng*', immediately before the start of section #D. Using Zhao's commentary it is again possible to restore the original diagram from its corrupt state. Whereas in (1) and (2) there seems little doubt that Zhao has supplied diagrams himself, the situation in this case is more problematic. Zhao's commentary, which lies between #D1 and #D2 clearly relates to the diagram: see appendix 3. It is then followed by #D2, which current editions present as main text:

> #D2 [46j]
> [Previously] in making this plan, a *zhang* 丈 has been taken as a *chi* 尺, a *chi* has been taken as a *cun* 寸, a *cun* has been taken as a *fen* 分, and a *fen* has been taken as 1000 *li* 里. [So overall] this used [a piece of] silk fabric 8 *chi* 1 *cun* square. Now a piece 4 *chi* 5 *fen* has been used, [so] each *fen* represents 2000 *li.*

If this really is part of the main text, the implications are interesting: somebody who copied the text before Zhao is telling us that he has reduced by half the scale of the diagram he found in the original – and even then the diagram was still about a metre across. Even if the text is really part of Zhao's commentary accidentally promoted into the main text (which could happen by starting the first column of a new page one character too high), it still must mean that he found the diagram already with the *Zhou bi*. The former interpretation is supported by the fact that #D2 is followed by these words presented as commentary:

[76] Qian produces five diagrams, all of which are in my judgement successful in representing mathematical situations discussed by Zhao. On the other hand, my reading of Zhao's text leads me to believe that Zhao himself only supplied one diagram, and that this was more or less identical with the first of those reproduced in later texts. For more detailed discussion, see appendix 1.

> 'Square' here refers to the diagram of [the boundary marking] the four poles – this takes to its limit all that is signified by the diagram of the seven *heng.*

This certainly does seem like the sort of thing Zhao might have written as a comment on the preceding text. The implication of all this is that despite the statement in Zhao's preface we cannot assume that all the diagrams in the text necessarily originated with him.

In another case it is evident that Zhao has actually excised part of the text as he found it, and replaced it by new material of his own. This is the shadow table of section #H, in which values of the noon shadow of an eight-*chi* gnomon for the twenty-four 'solar seasons' *qi* are given. As the section stands, the shadows are found by simple linear interpolation between the solstitial values used in the rest of the text. However, in his commentary to #H4 at the end of this section, Zhao tells us explicitly that the old table was at fault because it failed to make the equinoctial shadow values equal, and that therefore he has supplied new values. Fortunately he tells us enough about the faults of the old table for us to be able to reconstruct it: see appendix 3.

The very fact that Zhao tells us when he alters the text suggests two things. Firstly, it lessens the likelihood that he simply wrote the whole text himself and added his commentary as a blind to conceal the forgery. Secondly, it gives us some grounds for believing that unless he says otherwise he has passed on the text as he found it without conscious alteration or excision. This impression is strengthened when he complains that various pieces of material do not really belong with the rest of the text: see his commentary to sections #B1, #D3 and #K5. He has nevertheless left them undisturbed.

Zhen Luan

Zhen Luan 甄鸞 held office under the Northern Zhou 北周 dynasty (AD 557–581). He was responsible for the creation of the *Tianhe* 天和 system of mathematical astronomy, which was promulgated in AD 566. The usual heading in editions of the *Zhou bi* describe his role as having *chong shu* 重術 'reworked' the text. What he has done is in fact to find every instance in the main text or in Zhao's commentary where a calculation takes place and add detailed working. To illustrate this, let us consider an example. In section #B Chen Zi has explained a situation in which the observer on the flat earth is 60 000 *li* from the subsolar point, at which the sun is 80 000 *li* overhead. The text continues:

#B11 [26h]

... If we require the oblique distance [from our position] to the sun, take [the distance to] the subsolar point as the base, and take the height of the sun as the altitude. Square both base and altitude, add them and take the square root, which gives the oblique distance to the sun. The oblique distance to the sun from the position of the *bi* is 100 000 *li.* ...

Zhen's comment is as follows:

The minister Zhen states: The method for finding the oblique distance from [the observer's] gnomon to the position of the sun [is as follows]. First set out as base the distance of 60 000 *li* south to the point below the sun. Extend it again [i.e. set it out again on the counting board], and multiply it by itself, to obtain 3 600 000 000 as the base area. Further set out as altitude the sun's height of 80 000 *li.* Extend it again, and multiply by itself, to obtain 6 400 000 000 as the altitude area. Add the base and altitude areas to obtain 10 000 000 000 as the hypotenuse area. Take the square root, to obtain 100 000 *li* as the distance from the royal city to the sun ...

This is a typical example of the pedestrian nature of Zhen's contribution to the text. It is clearly unlikely to offer any strikingly new insights. The historian of mathematical education may note that the fact that Zhen thought his task worthwhile shows he was expecting the *Zhou bi* to be studied by readers with no more than basic arithmetical skills. None the less, he does assume that his readers can perform the four operations of arithmetic and extract square roots. Li Chunfeng 李淳風 , the third commentator on the text, points out that there are several errors in Zhen's illustrative calculations on Zhao Shuang's essay on the Pythagoras theorem: see pp. 18–22 in Qian's edition.

Li Chunfeng

As explained below (page 164), Li Chunfeng was responsible for editing the *Zhou bi* for inclusion in a collection of mathematical books prepared in AD 656. His comments principally relate to three areas:

(1) Zhao Shuang's essay on the Pythagoras theorem. As already mentioned, Li comments on and corrects Zhen Luan's illustrative calculations.

(2) After paragraph #B11, Li adds a long essay which makes two main points. Firstly, he criticises the fact that the calculations in the Zhou bi are based on a flat and parallel heaven and earth, despite the fact that section #E states that both heaven and earth bulge upwards. Secondly he points to experimental evidence that the 'one cun 寸 for a thousand li 里' shadow principle is false: see pp. 28–31 in Qian's edition.

(3) After Zhao Shuang's new shadow table in paragraphs #H2 and #H3, Li criticises the false assumption that noon shadow values throughout the year can be found by linear interpolation between solstitial values: see pp. 66–7 in Qian's edition. As he notes, hun tian 渾天 methods give results which contradict those of the gai tian 蓋天.

It will be clear therefore that Li's comments are those of an enlightened modern commenting on the technical deficiencies of his predecessors. While Li's work is of great interest in understanding the attitude of a seventh century astronomer and mathematician towards the *Zhou bi*, it is of little help in our primary task of understanding the book itself, and I shall therefore not discuss it further. Fortunately there is already a thorough discussion of this material by Fu Daiwie, which the interested reader may consult.[77]

The annotations of Li Ji

Further explanatory material is to be found in an appendix printed with the *Zhou bi* in the Southern Song print and a number of subsequent editions, though not reproduced by Qian. Its first appearance seems to have been in the Northern Song printing of AD 1084. This appendix was added by Li Ji 李籍, an official of the Song Imperial Library, and is in the traditional format known as a *yin yi* 音義, literally 'sound and sense'. Work of this kind is partly directed towards the fact that many Chinese characters can be read in two or sometimes more different ways, giving them both a different sound and a different sense. Li therefore goes through the text and lists problematic characters, for each one indicating the reading using the *fanqie* 反切 'turning and cutting' system, in which two well-known characters are given, one with the correct initial sound and one with the correct final sound. In some cases he adds more detailed explanation, none of which is of great interest in the present context.

Perhaps the only reason for mentioning Li Ji at all is that he enables us to feel slightly surer of the pronunciation of the title of the work we are studying. For the second character he gives the *fanqie* spelling '*bu mi qie* 步米切 meaning that we are to take the initial and final sounds of the characters *bu* and *mi*, yielding the reading *bi*. Of course the pronunciations I have given here are modern ones, and we have no real assurance of how the book's title was pronounced in the Song dynasty, let alone under the Han. But if we are to give the title a modern reading, we have some reason to prefer *Zhou bi* to the reading *Zhou bei* found in some other Western language studies.

[77] Fu Daiwie (1988).

The mathematics of the *Zhou bi*

From the Tang 唐 onwards the *Zhou bi* held a proud position as the first of the Ten Mathematical Classics. If however one was writing an old-fashioned history of Chinese mathematics based on the chronology of 'achievements', it would have to be said that the *Zhou bi* does not add a great deal to the credit balance of Han mathematics. There is no mathematical principle used in the *Zhou bi* that is not also well exemplified in another much more systematic Han mathematical treatise, the *Jiu zhang suan shu* 九章算術 'Nine Chapters on the Mathematical Art', which is the second of the Ten Classics. The *Zhou bi* is only unique in its use of the 'one *cun* for a thousand *li*' shadow principle to give the distances of celestial bodies.[78]

On the other hand, unlike the Nine Chapters the *Zhou bi* contains what may be called 'second-order' mathematical material in which a user of mathematics reflects on his activity. This occurs at the beginning of section #B, which I have called the 'Book of Chen Zi'. Rong Fang 榮方 opens the dialogue by asking for answers to a number of questions about the dimensions of the cosmos and the movements of celestial bodies. Chen Zi 陳子 assures him that mathematics can settle all such problems, but sends him away to find the answers for himself. Rong Fang twice returns to confess failure, and finally is favoured with the explanation he seeks. The pattern is effectively student:master :: suppliant:benefactor. A similar transaction between Confucius and a disciple is sketched in the opening of the *Xiao jing* 孝經 'Classic of Filial Piety', probably a work of Western Han date.

The most interesting part of this exchange comes in the words Chen Zi addresses to Rong Fang after the latter's first unsuccessful attempt to answer the problems he poses.

> #B6
>
> ... Now amongst the methods [which are included in] the Way, it is those which are concisely worded but of broad application which are the most illuminating of the categories of understanding. If one asks about one category, and applies [this knowledge] to a myriad affairs, one is said to know the Way. ... Therefore one studies similar methods in comparison with each other, and one examines similar affairs in comparison with each other. This is what makes the difference between stupid and intelligent scholars, between the worthy and the unworthy. Therefore, it is the ability to distinguish categories in order to unite categories *neng lei yi he lei* 能類以合類 which is the substance of how the worthy one's scholarly patrimony is pure, and of how he applies himself to the practice of understanding.. ...

78 If however we include the commentaries in our field of interest, it must be said that Zhao Shuang's essay on applications of the Pythagoras theorem is of considerable importance in the history of mathematics from any point of view: see appendix 1.

What we have here is a concise statement of a twofold heuristic strategy, summed up in the words 'distinguish categories in order to unite categories'. On the one hand one performs the analytic task of distinguishing different problem types from each other. On the other hand the very act of analysis brings together groups of similar problems which may be treated synthetically. Further, one can then attempt to 'unite categories' at a higher level by finding common structures underlying different problem categories.[79]

Chen Zi's analytic/synthetic approach is in fact not particularly well exemplified in the *Zhou bi* itself. It is however clearly the main rationale of the Nine Chapters. Whereas some (and only some) ancient Greek mathematicians were concerned to show how a great number of true propositions could be deduced from a small number of axioms, the author of the Nine Chapters followed a different but no less rational route in the reverse direction. He started from the almost infinite variety of possible problems, and aimed to show that those known to him could all be reduced to nine basic categories solvable by nine basic methods. To a great extent he succeeded, although the contents of some sections still show a degree of diversity. It did not strike him as worthwhile to try to argue explicitly that his methods would always work for the appropriate problem type. In the first place, he already knew they *did* work – the examples are before us to this day. Secondly, if it ever turned out that the method failed on a new problem, that would not have been taken as a sign that the method was wrong, but rather that it was necessary to distinguish a new problem category with a new common method for all problems of the new type – *lei yi he lei* 類以合類, in fact, as Chen Zi says.

Computational methods

The various sections of the *Zhou bi* differ in the amount of calculation they contain, but where calculations are found there are no radical differences in approach. In general it is simply assumed that the reader can add, subtract, multiply, divide, square and extract square roots using fairly large numbers with mixed units and fractional parts. At the time when the *Zhou bi* seems to have been assembled, such computations would have been carried out using counting-rod numerals arranged in matrix fashion.[80]

79 Chen Zi's insistence on the importance of *lei* 類 'categorisation' foreshadows later Chinese mathematicians' discussions of the heuristic importance of *bi lei* 比類 'comparing categories', on which see briefly Martzloff (1988), 154. In the preface to his commented edition of the Nine Chapters Liu Hui also makes it clear that for him the concept of *lei* is essential to his understanding of mathematics (Guo Shuchun edn, p. 177). I would incidentally suggest that no-one could read Kuhn (1970), 187–91 on how the student learns to apply theory without the sense that he is thinking on lines close to those of Chen Zi.

80 See Martzloff (1988), 194–6. Since the *Zhou bi* does not concern itself with the details of basic

If we focus on the text more sharply, however, differences of style emerge which reinforce the impression that the sections are not all by the same hand.

Thus after section #A, in which there is nothing that can be called computation apart from the observation that nine nines are eighty-one, we find that in section #B Chen Zi simply assumes that his student Rong Fang can cope with quite complex calculations without explanation. In one case this involves in effect finding the half-chord of a circle from the diameter and the length of the perpendicular from the centre to the chord: see #B29 and figure 8 on page 81 below. Thus Rong Fang has to recognise that he must apply Pythagoras' theorem, square two numbers, add and take the square root. Explanations of this process by Zhao Shuang and Zhen Luan occupy about 300 characters of commentary.

Further calculations follow rapidly. Near the beginning of section #D it is explained how the diameters of the successive *heng* 衡 circles may be obtained by adding a constant difference to preceding values. As each of these is obtained we are given the results of multiplication by three to obtain the circumference, and then dividing by 365 1/4 to obtain the length of one *du* 度, given in *li* 里, *bu* 步 and fractional parts. Once again, details are taken for granted.

Only at the end of the section do we find, in paragraphs #D20 and #D21, an attempt to follow a calculation in greater detail, preceded by the words *shu yue* 術曰 'the method says'.[81] The problem is to divide 119 000 *li* by 182 5/8. First we eliminate fractions by multiplying both numbers by 8 to obtain 952 000 *li* and 1461 as dividend and divisor. On division, the integral part of the quotient is in units of *li.* Since one *li* is 300 *bu*, the next step is to multiply the remainder by 3, and divide by 1461 to give the number of hundreds of *bu*. The remainder of this division is multiplied by ten, and then divided by 1461 to give the number of tens of *bu*. Again the remainder is multiplied by ten and divided by 1461 to give units of *bu*. The remainder from this division is simply set over 1461 as the denominator to yield a fraction of *bu.*

The calculation referred to here does not fit well into the overall pattern of section #D, whose final paragraphs seem to be in disorder. It may therefore be a note intruded by a later hand, or may have strayed in from another section. Section #E has no significant mathematical material, and section #F performs only simple calculations

arithmetical operations, it is hard to find any explicit reference to the material means of calculation in the main text. When a sequence of steps is detailed it is however usual for the procedure to commence with the instruction to 'set out' or 'align' *zhi* 置 the initial datum: see #D21, #G10, #G12, #G14, #H4, #I2, #I4, #I5, #I7, #I9, #I11, #I13, and #K14, #K15, #K16. Zhen Luan's commentary gives a little more detail: thus when a number is to be multiplied by itself it is first 'set up' *zhi*, and then we are told to 'extend it again and self multiply' *chong zhang zi cheng* 重張自乘 so that it is both multiplier and multiplicand. Compare Martzloff p. 204, which shows the arrangement of two such numbers in the 'upper and lower positions' on a counting board.

81 The precise significance of this phrase is unknown. Does it refer to an originally oral explanation of the text given by a teacher?

based on the shadow principle. Between #F6 and #F8, which are clearly continuous, #F7 has been intruded. Although it begins with 'the method says' it contains no details of calculations. This paragraph is unrelated to its context, and seems to be a displaced fuller version of #G2, which is clearly deficient in its present state.

It is not until we reach section #G, which appears to be in good order, that we find what is clearly a methodical attempt to explain the details of each calculation. Paragraphs #G10, #G12 and #G14 all give step-by-step instructions for division in the same way as #D21. Similar instructions are given in #H4 as part of the material accompanying the shadow table, although in this case we cannot be sure whether this may not be the work of Zhao Shuang rather than the original text.[82] Most of section #I consists of 'the method says' paragraphs detailing division procedures. The subject of section #J does not involve computation. Apart from the fragmentary and possibly intruded paragraph #K6, there is nothing resembling 'method' details to accompany the division computations in section #K, with which the *Zhou bi* ends. About the only interest of the 'method' paragraphs are that they enable us to say something about the mathematical level assumed in their audience. It seems that the reader of the sections in which they occur was expected to be able to perform straightforward long division using quite large whole numbers, presumably on a counting board, but that he might require help in dealing with mixed units and fractional parts. As I have already mentioned, the writer of section #B seems to be aiming at a rather more sophisticated readership for whom such help would not be necessary.

Geometry

Similar triangles?

It has been common to speak of the *Zhou bi* as using the principle of similar triangles in its calculations about the height and distance of heavenly bodies. Until I came to write this section, I followed the same practice, and it is certain that the *Zhou bi* does discuss problems of a type that a modern mathematician would classify in this way. But as I re-read the relevant parts of the *Zhou bi* I became uncomfortable about continuing to speak like this.

The problem is that at no point does the text compare one triangle with another, and indeed it contains no noun corresponding to our 'triangle' at all. Plane figures bounded by three straight lines just do not figure as a unit of discourse. There are separate terms for what we would now call the base, altitude and hypotenuse of a right-angled triangle – *gou* 勾 ('hook'), *gu* 股 ('thigh') and *xian* 弦 ('bowstring'). However, rather than naming the ensemble of all three, the *Zhou bi* focuses on the

[82] See appendix 4.

combination of *gou* and *gu* to make the carpenter's L-shaped trysquare *ju* 矩.[83] This is frequently referred to in section #A, and is also mentioned in the fragmentary section #C. If we are concerned to penetrate the thinking of ancient mathematicians rather than to force our own categories upon them, it seems that the term 'similar triangles' would be better avoided in discussing the *Zhou bi*.

How then does the *Zhou bi* deal with the (now temporarily anonymous) genre of problems in question? A careful reading of part of section #B is enlightening. In response to Rong Fang's importunity, Chen Zi has just revealed the shadow principle to him – i.e. that for every 1000 *li* further south towards the noon sun, the shadow of an eight-*chi* gnomon diminishes by one *cun*. He continues:

> #B11 [26h]
> 'Wait until the base is six *chi*, then take a bamboo [tube] of diameter one *cun*, and of length eight *chi*. Catch the light [down the tube] and observe it: the bore exactly covers the sun, and the sun fits into the bore. Thus it can be seen that an amount of eighty *cun* gives one *cun* of diameter. So start from the base, and take the *bi* as the altitude. 60 000 *li* from the *bi*, at the subsolar point a *bi* casts no shadow. From this point up to the sun is 80 000 *li*. If we require the oblique distance [from our position] to the sun, take [the distance to] the subsolar point as the base, and take the height of the sun as the altitude. Square both base and altitude, add them and take the square root, which gives the oblique distance to the sun. The oblique distance to the sun from the position of the *bi* is 100 000 *li*. Working things out in proportion, eighty *li* gives one *li* of diameter, thus 100 000 *li* gives 1250 *li* of diameter. So we can state that the diameter of the sun is 1250 *li*.'

The situation is as in figure 7, but see below as to the caution needed in using such an illustration. To follow Chen Zi's mental processes, we need to attend closely to his order of discourse. 'The base' here is the noon shadow of the eight-*chi* gnomon, fairly clearly selected to give the convenient 3:4:5 ratio between base, altitude and hypotenuse. There then follow the two sentences about the eight-*chi* sighting tube. The next two sentences 'So start ... 80 000 *li*' are the key. 'The base' is the 60 000 *li* from the gnomon to the subsolar point. Since 6 *chi* of shadow gives 8 *chi* of gnomon height, the simple numbers involved enable an intuitive leap to be made to the realisation that the height of heaven is 80 000 *li*. We tend to assume in such a situation that we are dealing with something like the ratio equivalence:

[83] Zhao Shuang uses *ju* to refer to the L-shaped areas in his Pythagorean essay, although he has no special word for the triangular areas he discusses, which are just 'the red areas *zhu shi* 朱實 [on the diagram]': see Qian's edition, p. 18. While I am aware that an L-shape may be referred to as a 'gnomon' in Western mathematics, I avoid the term here to prevent confusion with the vertical pole designated by the same term in my translation of the *Zhou bi*.

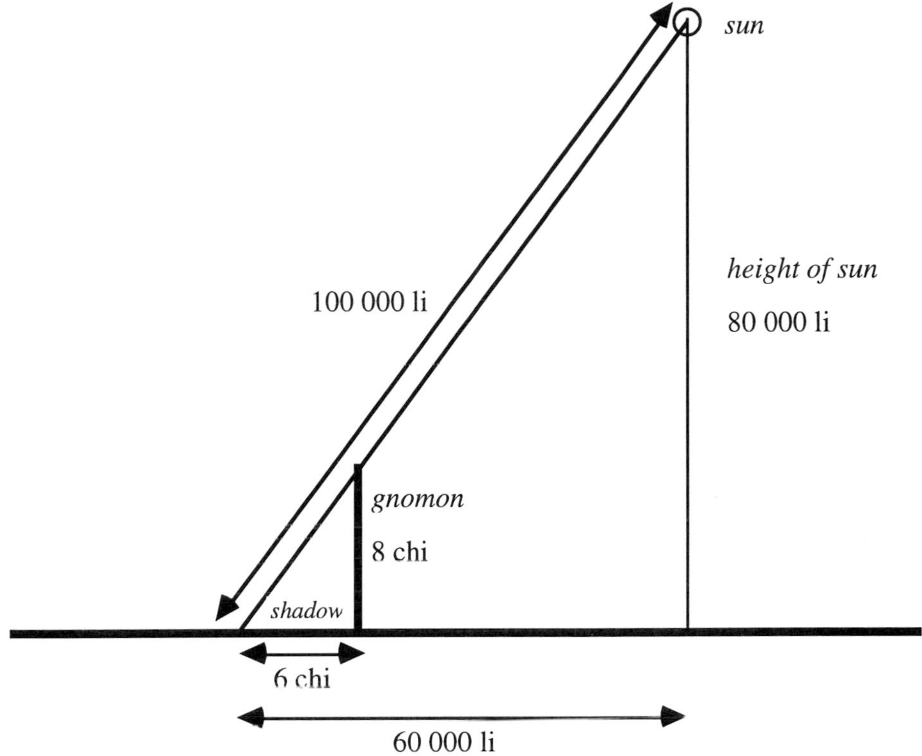

Figure 7. Chen Zi's sun sight.

height of heaven = 80 000 *li*.

(6 *chi*) : (8 *chi*) :: (60 000 *li*) : (height of heaven)

which leads on to the equation:

height of heaven = 60 000 *li* × (8 *chi*) ÷ (6 *chi*), so that:

height of heaven = 80 000 *li*.

In fact I suggest we could equally well be dealing with a less formally mathematical pattern of thought, equivalent to the paradigm/syntagm pattern applied by Saussure to linguistic analysis:[84]

[84] Saussure (1983). For the use of this concept in the field of early Chinese thought, and more generally,

	base	altitude	Paradigm
(gnomon)	6 *chi*	8 *chi*	
(heaven)	60 000 *li*	? *li*	

Syntagm

From the complete syntagm for the base, we can make the immediate leap to see that the missing quantity for the altitude must be 80 000 *li,* since the pattern is clearly one in which the dimension *chi* can be replaced by 10 000 *li.* The process is more verbal than computational. It is perhaps an example of what Chen Zi called 'uniting categories'. A similar pattern seems to underlie the calculation of the solar diameter. We move from the fact that 'eighty *cun* gives one *cun* of diameter' to the fact that 'eighty *li* gives one *li* of diameter' by simply replacing *cun* by *li.* Only when we have to find how many *li* of diameter we get for 100 000 *li* do we need to calculate, and clearly all that is being done is to find how many lots of 80 *li* there are in 100 000 *li.* We can only call such reasoning 'an application of the principle of similar triangles' in a purely Pickwickian sense.[85] To illustrate it by a carefully labelled Euclidean diagram when none is referred to in the text is perhaps only a way of misleading oneself further about what Chen Zi is really up to.

Despite the way the *Zhou bi* is commonly characterised, the passage we have just analysed is the only one in which this type of problem is discussed. In #E7 the height of heaven above earth is stated to be 80 000 *li,* but no calculation is involved. In the detailed discussions of stellar observations using the gnomon in section #F, it is simply taken for granted that one can move from measurements at the gnomon to measurements on the heavens by replacing each *chi* by 10 000 *li* as in the paradigm set out above. On the face of it, nothing worthy of being called computation is involved. We may of course wish to believe that behind these statements in the *Zhou bi* there were calculations that 'justified the deduction properly' in our terms. But the fact remains that there is no evidence that this was the case, or that the authors of the text cared about such matters. In such circumstances, it is surely better history to take the ancient author at his word when it comes to accounting for his thought processes.[86]

see Graham (1986), 16ff.

[85] If any kind of similarity is involved, it is the similarity of 'trysquares' formed of base and altitude rather than of full triangles. Notice how we move from the base and altitude on a small scale to calculate the base and altitude on the celestial scale, then use these two last quantities to calculate the required celestial hypotenuse. We do not first calculate the small scale hypotenuse and then move to the larger one by proportion, which might have seemed more natural if the similarity of entire triangles had been at the front of the author's mind.

[86] For comparison, see problem 22 of chapter 9 of the *Jiu zhang suan shu* 九章算術, which deals with a

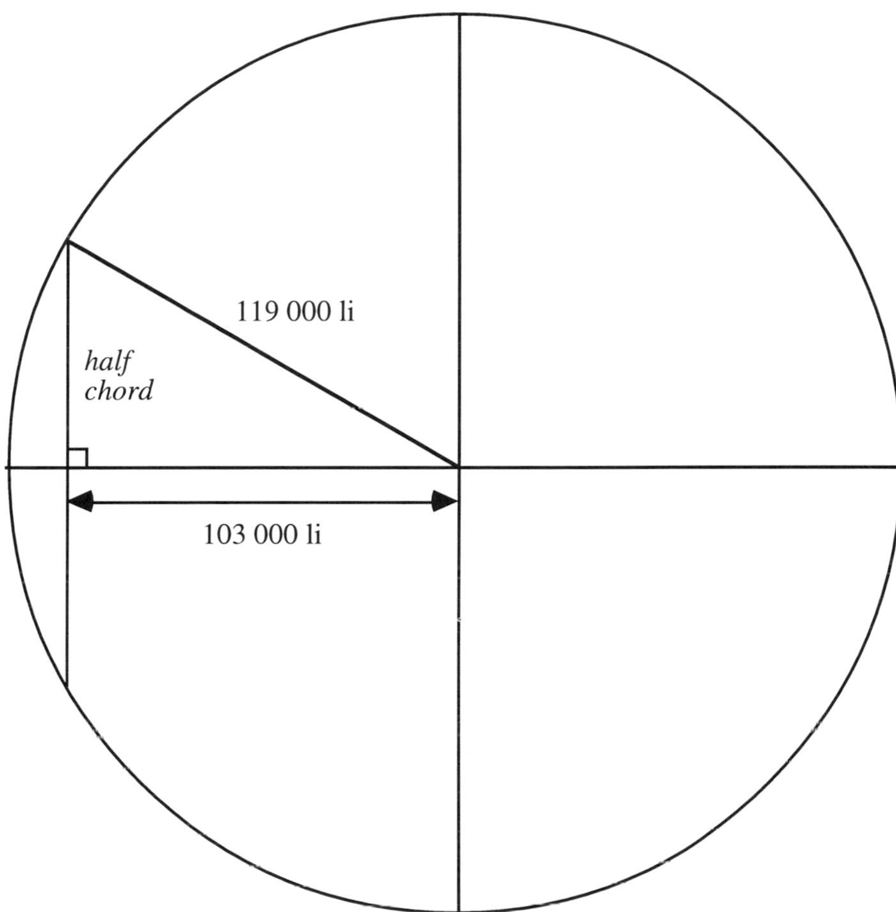

Figure 8. The Pythagorean relation applied to find the half chord as in #B29.

The Pythagorean relation

Certain calculations in the main body of the *Zhou bi* assume that in a right-angled triangle:

terrestrial sighting calculation similar to that in the *Zhou bi*, except that the shadow principle is not used. A man of known eye height is sighting over a tree of known height a known distance away on the top of a mountain at a known distance but unknown height. Labelling these quantities m, t, d, L and h respectively, the text obtains the mountain's height by performing a computation equivalent to:

$$h = (t - m) \times L / d + t.$$

The text gives us no indication of its author's mental processes The problem may shout 'similar triangles' to us, but we have no right to assume that the ancient Chinese mathematician thought that way, and as we have seen there is good reason to think that he did not.

$$(\text{base})^2 + (\text{altitude})^2 = (\text{hypotenuse})^2.$$

The first example met with is in paragraph #B11, which has just been discussed; from a base and altitude of 60 000 *li* and 80 000 *li* a hypotenuse of 100 000 *li* is determined. Two more difficult applications are met in paragraph #B29, and there is a similar one in #B33. Taking the first as an example, we are in the situation of figure 8: a chord of a circle of radius 119 000 *li* has a perpendicular distance of 103 000 *li* from the centre. It is required to find the half chord. We would naturally calculate thus:

$$(103\,000\ li)^2 + (\text{half chord})^2 = (119\,000\ li)^2.$$

Leading to the result:

half chord = 59 598.66 *li* to two decimal places.

It is however interesting that in his commentary Zhao Shuang proceeds slightly differently. He works with double the values given in the above equation, so that the whole chord length rather than the half length is the result after taking the square root. This he gives as 119 197 *li* 'and a bit' *you ji* 有幾. When halved (the 'bit' evidently being ignored) this yields the result given in the *Zhou bi* main text, which is 59 598 1/2 *li.* The doubling, which seems to have no geometrical significance, is presumably a device to ensure a reasonably accurate value for the root without having to carry the extraction into fractions.[87]

Apart from these examples, the *Zhou bi* contains no other applications of the Pythagorean relation to find an unknown side of a right-angled triangle.

The only remaining material relating to this topic is found in section #A. It has in my view attracted a degree of exegetical attention somewhat out of proportion to its significance, since it seems to amount to no more than a garbled reference to the case of what we would call a 3–4–5 triangle. There was some excuse for this overemphasis when traditional scholars were still convinced that section #A was the ancient core of the whole book, preserving a dialogue between the Duke of Zhou and a sage of the Shang dynasty around 1000 BC. Qian Baocong attacked this view over sixty years ago, and I shall argue later that it may even be the case that the present opening section of the *Zhou bi* was the last to be added, and was perhaps intended more to impress its readers than to train them in mere practical techniques: see page 153ff.

It is however difficult to pass lightly over such well-trodden ground. The two relevant paragraphs are therefore discussed here. To illustrate the difficulties faced by the first interpreter of the text, Zhao's comments have been added after relevant sections of the main text, which are printed in bold type for clarity, together with the original Chinese.

[87] In his additional commentary, Zhen Luan continues heroically until he has calculated the fractional part of the final result as (1/2) + (75 191 /476 790).

#A2 [13f]

(a) Shang Gao replied 'The patterns for these numbers come from the circle and the square 商高曰 數之法出於圓方.

Zhao comments: 'When the diameter of a circle is one, its circumference is three. When the diameter of a square is one, its perimeter is four. Stretch out the circumference of the circle to make the base, and develop the perimeter of the square to make the altitude, so that together they make a corner, the inclined line across which makes the hypotenuse of five. These are the mutual proportions of the circle, the square and the inclined diameter. Thus [the text] says "The methods for these numbers come from the circle and the square." The circle and square are the shapes of heaven and earth, and [embody] the numbers of Yin and Yang. So since what the Duke of Zhou asked about was heaven and earth, Shang Gao set out the shapes of circle and square to manifest their counterparts, and based himself on odd and even numbers to regulate their patterns. This is called speaking briefly with a far-reaching purpose, minutely subtle and darkly penetrating.'

(b) The circle comes from the square, the square comes from the trysquare 圓出於方方出於矩,

Zhao comments: 'The numbers of the circle and the compass are given order by the square. The square [here refers to its] perimeter [of four]. Things that are square and straight come from the trysquare. The trysquare is breadth and length.'

(c) and the trysquare comes from [the fact that] nine nines are eighty-one 矩出於九九八十一.

Zhao comments: 'Extrapolating the proportions of circle and square, penetrating the numbers of breadth and length, you have to multiply and divide in order to do the reckoning. Now 'nine nines' [i.e. the multiplication table, which begins with these words] is the source of multiplication and division.'

Taking Zhao's comments for the whole paragraph together, he clearly faces much the same problems as we do in making sense of the text. By pointing out the connection between the circle and square and their corresponding numbers he does however help to link this paragraph with the next one, in which three and four will appear as the base and altitude. A first century BC parallel comes from the *Shuo yuan* of Liu Xiang:

Therefore the sage in his sagehood is like the carpenter's square circuiting in four, like the compasses circuiting in three, when the round is completed they begin again ... (*Shuo yuan* 19, 2b, *Sibu congkan* edn)[88]

[88] I owe this reference to Graham (1978), 308–9, whose translation I follow here. Graham is discussing

#A3 [14b]

(a) 'Therefore fold a trysquare 故折矩

Zhao comments: ' "Therefore" here is a word used to indicate the further development of a subject [i.e. there is no strictly deductive relation with the preceding text]. He is going to deal with the proportions of base and altitude, so he says "fold a trysquare."'

(b) so that the base is three in breadth 以為勾廣三,

Zhao comments: 'This corresponds to the circumference of a circle. That which runs crosswise is called "broad', and thus the base is "broad". "Broad" [here refers to] the short [side].[89]

(c) the altitude is four in extension 股脩四

Zhao comments: 'This corresponds to the perimeter of the square. That which is lengthwise is called "an extension", thus the altitude is "an extension". "Extension" [here refers to] the long [side].'

(d) and the diameter is five aslant 徑隅五.[90]

Zhao comments: '[These are] the natural proportions in correspondence with each other. "Diameter" here means "straight" and "aslant" here means "corner". It is [what is] also called the hypotenuse, *xian*.'

(e) Having squared its outside, halve it [to obtain] one trysquare 既方其外半之一矩.[91]

the definitions A58 and A59 of the Mohist Canon, which refer to the compasses and trysquare. A59 is corrupt, but is tentatively rendered by Graham as '*Fang* (square) is circuiting in four from a right angle'.

89 It was a convention that the base should be shorter than the altitude: see for instance the careful explanation given by Liu Hui when the *Jiu zhang suan shu* departs from this practice: 9, 420–1 (fifth problem of chapter) in the edition of Guo Shuchun, (1990).

90 It is interesting that the main text of this section does not use the otherwise universal term *xian* 弦 'bowstring' for the hypotenuse.

91 This follows the emended version in Qian's edition. The Song text (1, 2a), followed by all other traditional texts, is even more cryptic, if not ungrammatical: *ji fang zhi wai ban qi yi ju.* 既方之外半其一矩. A tentative rendering would be 'When [..?..] the outside of the square, halve one of its trysquares'. However,, when Zhao's commentary on this passage recites the text immediately afterwards, he has *ji fang qi wai* [...] *ban qi yi ju.* 既方其外 [...] 半其一矩. 'Having squared its outside [...] halve one of its trysquares'. . In his text prepared for the *Siku quanshu* Dai Zhen stated that he could make nothing of the received text, and proposed to read *ji fang wai wai ban zhi yi ju* 既方外外半之一矩 'Once outside the square, halve it outside [..?..] one trysquare'. I do not find this very helpful. The explanation given in Dai's note to this section does little more than recite the facts of the Pythagorean relations between the sides of a 3–4–5 triangle in a somewhat confused manner: see vol. 786, p. 5 in *Siku quanshu* photographic reprint. Following the 武英殿 'Palace Edition', which is based on Dai Zhen's text with some modifications, Qian

Zhao comments: 'The pattern in base–altitude [problems] is that you first know two numbers, and then deduce one [more]. If you see the base and altitude, then you seek for the hypotenuse. You first multiply each by itself, and form the area. When the area is formed, then the circumstances change and manipulations proceed. Therefore [the text] says "having squared its outside". One may either add the base and altitude areas in order to seek the hypotenuse, [or] in the midst of the hypotenuse area one may seek the separation and addition of the base and altitude [areas]. These areas are not exactly the same, so one may proceed to take from and give to them, and [so that] each has something to receive. Therefore [the text] says "halve it [to make] one trysquare". The method is: multiply both base and altitude by themselves, so three threes are nine and four fours are sixteen. Add to make twenty-five, the area from multiplying the hypotenuse by itself. Subtract the base [area] from the hypotenuse [area], and one gets the altitude area of sixteen. Subtract the altitude [area] from the hypotenuse [area] and you get the base area of nine.'

While this comment does not add very much to our understanding of the text, it does lay out the basic program to be followed in Zhao's subsequent essay. If the *Zhou bi* text ever meant anything coherent, it seems likely that it is in some way connected with the fact that if the square on (say) the base is inscribed in one corner of the square on the hypotenuse, then the remaining trysquare-shaped area is equal to the square on the other side: for further discussion see appendix 1.

(f) Placing them round together in a ring, one can form three, four and five 環而共盤得成三四五.

Zhao comments: '*Pan* 盤 is to be read as *pan* in [the expression] *pan huan* 盤桓. This means take the accumulation of sums and differences, place them round together bent into a ring, take the square root, and you get one side. Therefore [the text] says: "you can form three, four and five."'

The only relevant sense of *pan huan* amongst the five given by Morohashi's dictionary, 23568.55, seems to be the last 'a woman's chignon'. The reference is presumably to the way the trysquares are formed into a ring like a woman's hairwinding round. In his Pythagorean essay Zhao does discuss a situation in which the base area

exchanges *zhi* and *qi*. in the Song version of the main text to produce the version given here. It is not easy to find a plausible emendation of the text that conveys much meaning. If it is allowable to suggest that 外 may be graphically corrupted from the left hand side of 裡, and that 半 may be the remains of the top of 卷, then we have 既方其裡卷之一矩 'when you have placed a square within, what wraps it round is a trysquare'. The expressions here can be paralleled from Zhao Shuang and Liu Hui, and the sentence refers clearly to the kind of manipulation of areas discussed in Zhao's essay (appendix 1).

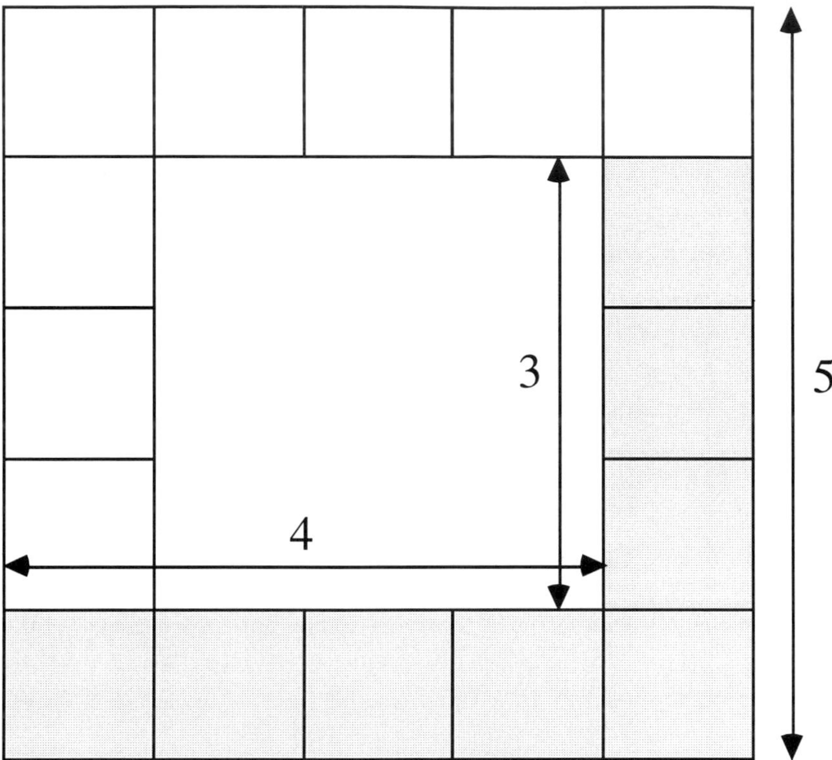

Figure 9. Trysquares in a ring.

and altitude area trysquares are formed into what might be called a ring, and the resulting manipulations do in fact lead to finding the three sides: see appendix 1, figure 27. Whether that is what this section of the *Zhou bi* meant is another matter. It may simply be that we are supposed to take two trysquares as in figure 9 and form a square of side five, in which case we can fairly readily identify the dimensions of three, four and five units in the diagram, and the area of 25 squares which results may possibly be the 'combined length' referred to below.[92] But unlike traditional Chinese

[92] The interested reader may refer to two other attempts to make sense from this passage. Cheng Yaotian 程瑤田 (1725–1814) in his *Zhou bi jushu tuzhu* 周髀矩數圖注 1b–2a believes that what are 'placed in a ring' are the two sides of a trysquare whose inner edges are 3 and 4, and a length 5 long. While this is convincingly simple, in my opinion he does not succeed in making it clear how this idea relates to the preceding text. Feng Jing 馮經 (fl. 1750) is similarly straining for a meaning: *Zhou bi suan jing shu* 周髀算經術, 2.

scholars we are not obliged to interpret our puzzlement as an inability to comprehend the thought of some ancient genius. It may simply be that what we have before us is simply a garbled and pretentious statement of the fairly obvious.

(g) The two trysquares have a combined length of twenty-five. This is called the accumulation of trysquares 兩矩共長二十有五是為積矩.

> Zhao comments: 'The "two trysquares" are the areas [formed by] multiplying the base and altitude by themselves. The "total length" is the number of the sum of the areas. [The text] is about to apply [this relation] to a myriad things, so here it first sets out the proportions.'

Once again Zhao's explanation accords with his essay given in appendix 1.

(h) Thus we see that what made it possible for Yu to set the realm in order was what these numbers engender' 故禹之所以治天下者此數之所生也.'

> Zhao comments '[In remote antiquity] Yu controlled the Great Flood, and drained it away along the [Yangzi] Jiang and the [Yellow] River. He [had to] survey the shapes of the mountains and waterways and determine the circumstances of height and depth, in order to eliminate the disaster afflicting the empire, and to lift away the distress. He made [the floods] pour eastwards into the sea so that they did not return. It was in this [task] that the [methods of] base and altitude originated.'

For the story of Yu's work, see the *Yu gong* 禹貢 'Tribute of Yu' chapter of the Book of Documents.

Before leaving this topic I must deal with the commonly made claim that this section embodies something like a proof of the Pythagoras' theorem.[93] I think it should be clear from my translation that there is nothing in the main text that could be considered an attempt at a proof, unless we make a major effort of the imagination followed by heroic emendation. Nor is there anything in Zhao Shuang's commentary to suggest that he can see a proof in this material: see appendix 1.

So far as I can tell, the crucial error behind all attempts to find a proof here is the mistaken belief that the first of the three diagrams found in all editions, the *xian tu* 弦圖 'hypotenuse diagram', belongs with the main text of section #A. This is clearly wrong, for three reasons:

[93] See for instance Needham (1959), 22–3 and 95–6, followed by van der Waerden (1983), 26–7, also by Lam and Shen (1984), 88–9. Needham's translation of the relevant material follows an interpretation said to be due to Arnold Koslow, in which the text is made comprehensible by rewriting it with major additions. Similar ideas are put forward in Swetz and Kao (1977).

(1) Zhao's states in his preface that he 'added diagrams in accordance with the text'. This implies clearly that there were no diagrams before he put them there.
(2) The main text of #A makes no mention of any diagram at all.[94]
(3) On the other hand in his subsequent essay Zhao makes explicit and full use of this diagram for the purpose of explaining his manipulations of base, altitude and hypotenuse. It is in fact the only diagram he mentions, and the other two seem to be later additions by someone who did not understand Zhao's work. Throughout his discussion Pythagoras' theorem is used, but no attempt is made to justify it.

Of course I do not deny that with sufficient ingenuity you can use the 'hypotenuse diagram' to give a general proof of the Pythagoras theorem by the dissection and re-arrangement of areas.[95] The point is that neither the main text nor Zhao do in fact use it for this purpose.

In fact the only evidence we have for anything that could be called an ancient Chinese proof of the Pythagoras theorem appears in the third century commentary of Liu Hui 劉徽 to the *Jiu zhang suan shu* 九章算數, near the beginning of chapter nine, *Gou gu* 勾股 'Base and Altitude', which uses the relationship repeatedly. Liu is commenting on the first three problems of the chapter, which use the case of base 3, altitude 4 and hypotenuse 5 to illustrate how each side may be found from the other two. In particular, he explains the instruction to find the hypotenuse from the square root of the sum of the squares of the other two sides:

> You multiply the base by itself to make the red square, and multiply the altitude by itself to make the green square. Then you make the areas go in and out and build each other up *chu ru xiang bu* 出入相補, each following its kind, and so get to a state where the surplus does not shift or move, where together they form the superficies of the hypotenuse square. Divide this by opening the square, and this is the hypotenuse. (*Jiu zhang suan shu*, 9, 419, Guo Shuchun edn)

All Liu Hui's original diagrams were lost by the Song, and any attempt at restoration must be conjectural – but see the ingenious attempt of Wagner (1985). It does not seem that he is using the 'hypotenuse diagram' in which the areas coloured red were triangular. The point however is that, whatever diagram he was using, Liu sees no point in giving the reader more than a vague gesture at the process to be carried out, after which he continues with his business of explaining the main text. The whole

[94] This point is also made by Nōda (1933), 17.

[95] This is done quite elegantly by Lam and Shen (1984), 88–9. Even then there are major difficulties. The procedure they describe is clearly not mentioned in the main text, and not even in the rewritten version of Koslow which they quote from Needham. Moreover, their proof does not require the whole of the 'hypotenuse diagram'.

affair is worth no more than a passing mention – there are certainly no signs of hecatombs of oxen meeting their end in commemoration. And that is the main point: even when an ancient Chinese mathematician gives a proof, it is not very important to him in comparison with his real aim of explaining the use of the methods he is expounding to solve specific problems.[96]

Why should it be otherwise? When one reads certain writers on this subject it is clear that they feel that the work of China's ancient mathematicians is in some way second class if they cannot be shown to have 'had proofs' of the methods they used. But as Lloyd has argued strongly, the ancient Greek obsession with the impossible demand for ultimate justification was by no means a wholly positive phenomenon, and it depended for its appearance on very specific social and historical conditions.[97] Why it should be thought either likely or meritorious for Chinese mathematicians to have felt the same way is not obvious.

One problem in interpreting the material of section #A is that it is difficult to find any close parallels in other ancient texts. There are however some examples which I do not think have so far been much noticed by scholars. The first relates to the odd expression used in (a) above, where we are told to form the base of three and altitude of four by 'folding a trysquare' *zhe ju* 折矩.

In AD 5 Wang Mang held a great conference of experts in 'lost classics, ancient records, astrology, calendrical calculations, [and the metrology of] bells and pitchpipes'. A summary of the proceedings of this meeting was written by Liu Xin, and is now contained in chapter 21a of the *Han shu.*[98] There we are told of the cosmic significance of the steelyard *heng* 衡 and its weights *quan* 權:

> As for the steelyard and its weights, the steelyard represents what is level, and its weights represent what is heavy. The steelyard bears the weights and equalises things, levelling out the light and heavy. Its way is that of balance. It makes manifest the trueness of the water-level, and the straightness of the plumbline. Turned to the left it makes manifest the compasses; *folded to the right it makes manifest the trysquare* 左旋見規右折見矩. In the heavens it assists the *xuan ji* 旋機[99], and gives consideration to the indication of the establishments [of the months] in order to regulate the seven [processes of] government, and so it is called the *yu heng* 'jade steelyard'.[100] (*Han shu* 21a, 969)

[96] Chemla (1991) investigates this and related questions, arguing that the expression of an algorithm may be structured so as to hint at its proof.

[97] Lloyd (1990), chapters 3 and 4.

[98] See *Han shu* 12, 359 and 21a, 955, where Ban Gu warns us that he has censored Liu's account.

[99] For the significance of this term within the *Zhou bi*, and references to the wider context, see the discussion later in this chapter.

[100] This expression, on which see Cullen and Farrer (1983), 55ff. and Cullen (1981), 39–40, is evidently

It is hard to make much sense of the references to right and left in physical terms. Cosmologically speaking the matter is clearer, since we know that left is Yang and right is Yin: following Zhao's commentary on (a) we may draw up the following list of correspondences:

Yin	Yang
Earth	Heaven
right	left
even	odd
four	three
square	circle
trysquare	compasses

As for giving any physical meaning to 'folding' all I can think of is some sort of portable balance arm that folds up in the middle. This could certainly be used to form a trysquare shape if it was 'folded' to a right angle, and from another point of view the two parts of the device could be viewed as the two legs of a compass. It is of course not the fault of Liu Xin that what seemed to him a deep insight into the formal structure of the cosmos now seems so trivial. For our present purpose, however, the point is that the obscure phrases of Shang Gao can be related to the way a known group of people talked at a particular time. It is also possible to produce close parallels from the same source to Shang Gao's earlier references to the circle and the square, although the relation between them seems to be reversed:

> Weights are balanced against things and produce the steelyard; the steelyard turns and produces the compasses; the compasses [make a] circle and produce the trysquare; the trysquare [makes a] square and produces the plumbline; the plumbline [makes a] straight line and produces the water-level. When the water-level is true it levels the steelyard and balances the weights [and the cycle begins again]. ... the compasses and trysquare require each other, and the positions of Yin and Yang are in order, so that the circle and square are formed. (*Han shu* 21a, 970)

The significance of these parallels will be discussed further below. Another possible parallel to the obscure wording of section #A comes from the *Kao gong ji* 考工記 'Artificer's Record' section of the *Zhou li* 周禮. Once more there is a link with Wang Mang, since he claimed to be following the pattern of state organisation laid down in this book, which was traditionally thought to have been written by the Duke of Zhou, Wang Mang's chosen role model. It will be remembered that section #A contains the

used here as a name for the Northern Dipper, idealised as pointing to each of the 12 directions corresponding to the 12 earthly branches at dusk during the 12 months of the year. These are the 'establishments' of the months in question.

puzzling phrase *ban qi yi ju* 半其一矩 , literally 'halve its one trysquare'. Following Qian we emended *qi* 其 to *zhi* 之, giving the only slightly less mysterious 'halve it [to obtain] one trysquare', which at least avoids us having to imagine what half a trysquare might mean. But in the *Zhou li* we find a description of the making of L-shaped stone chimes which reads as follows:

> The Chime Master makes [stone] chimes. Their bent hook is one trysquare and a half, *yi ju you ban* 一矩有半. (*Zhou li zhu shu (Kao gong ji)* 41, 8b.)

Shortly after this we find:

> In the tasks of the Cart Men half a trysquare is called a *xuan* 宣. One *xuan* and a half is called a *shu* 欘. One *shu* and a half is called a *ke* 柯. One *ke* and a half is called a *qing zhe* 磬折 'chime fold'. (*Zhou li zhu shu (Kao gong ji)* 41, 9a–10a.)

From the second passage it is evident that:

$$
\begin{aligned}
1 \text{ 'chime fold'} &= 1 \text{ 'trysquare'} \times (1/2) \times (3/2) \times (3/2) \times (3/2) \\
&= 1 \text{ 'trysquare'} \times (27/16)
\end{aligned}
$$

which is approximately one and a half trysquares as in the first passage. But of course the main problem is that we have no idea what it means to multiply a trysquare by some number, let alone a fractional number. All we can say for the present is that it may well be that section #A is not so much corrupt at this point as referring to some process we do not understand.[101]

Circles and angular measure

Although circles are mentioned very frequently throughout the *Zhou bi* (see for example section #D), the work uses only the most elementary geometrical concepts relating to them. Except for the graduated circle on flat ground described in section #G and the fragment in #C, all the circles discussed centre on the north celestial pole. We are told of their diameter, *jing* 徑 and circumference, *zhou* 周. Whenever these two quantities are to be related (as in the first instance in #B16) it is assumed that they are in the ratio 1:3. This value is also used in the *Jiu zhang suan shu*: the first known Chinese efforts

[101] In his *Kao gong ji tu* 考工記圖 'Illustrations to the Artificers' Record' Dai Zhen attempts to explain this passage without obvious success: 2, 12a–13b in *Anhui congshu* edition. Another attempt is made by Dai Zhen's contemporary Cheng Yaotian 程瑤田, who is convinced that the text is talking about angles, with a 'trysquare' meaning 90°: see his monograph *Qing zhe gu yi* 磬折古義 'On the ancient meaning of the expression "chime fold"', also in *Anhui congshu*. On the other hand the Eastern Han commentary takes all the units mentioned as measures of length, which is supported by the fact that in the next section a *ke* is said to be three *chi* long, which would make the 'trysquare' 2 2/3 *chi*. Fortunately this question is not one which need detain us unduly.

to obtain a more accurate value date from the first and second centuries AD.[102] As already mentioned, there are three instances of the calculation of a chord of a circle. The nature of the *Zhou bi*'s concerns means that there are no discussions of problems involving circular areas.

It is in the context of a discussion of circles in the *Zhou bi* that the question of angular measure arises. Once again, I suggest that we need to tread very carefully to avoid interpreting the thought-patterns of ancient authors in terms of our modern preconceptions. I use the term 'angular measure' here for want of a better, but from the outset I must state my conviction that the concept of angle as found (say) in Euclid is wholly absent from early Chinese mathematics. Now since the word *du* 度, nowadays translated as 'degree' is found frequently throughout the book (e.g. in sections #D, #G, #I and #K) this sounds perverse. Admittedly the *Zhou bi* has 365 1/4 *du* to the complete revolution rather than 360, but a mere difference of calibration hardly seems enough to banish the concept of angle altogether.

There is however a genuine distinction to be made here. I cannot now discuss the origin of the angle-concept in the ancient West, but to anyone reading Euclid (*c*. 300 BC) it is clear that by his time an angle had become a generalised measure of the rotation of one line relative to another in the same plane. Thus wherever two line segments intersect, one naturally focuses attention on the angles formed. Nothing of the kind occurs in early Chinese mathematics as evidenced in the *Zhou bi*, the *Jiu zhang suan shu*, and indeed all the ancient mathematical classics. In all the complex geometrical problems they discuss, the focus is always on the values and ratios of lengths. The *du* is never used as a measure of the angle between two lines, or as a graduation for a terrestrial circle – unless (as in section #G) that circle is a model of a celestial circle.

The *du*, then, is wholly confined to the heavens in ancient China. As we have seen, it certainly had its origin there, either as a unit of the sun's daily motion or possibly as a measure of the interval in days between the dusk transits of successive celestial bodies. The same could of course be said of the Western degree: why this measure soon descended to earth while its cousin remained in orbit must remain a matter for speculation.[103]

The calendrical astronomy of the *Zhou bi*

The interest of the *Zhou bi* in terms of calendrical astronomy is not that it tells us anything totally unexpected. Its great value is that it is the only extended piece of

[102] See Needham (1959), 99–102 for a brief survey with references to primary sources.

[103] On the origin of the sexagesimal division of the circle in the West, see Neugebauer (1975) vol. 2, 589ff.

unofficial writing on this topic surviving from the Han period.[104] The content and form of its calendrical material therefore deserve careful study.

Organisation of calendrical material

From the *Han shu* onwards, the description of a system of calendrical astronomy in the standard histories *zheng shi* 正史 takes on a fairly stereotyped form. The text begins with a list of numerical constants, each given a name indicative of its function within the system. Then follow a series of 'procedure texts', in which it is explained how the fundamental constants of the system can be used to predict astronomical phenomena. As already mentioned, an important part of such systems was the 'system origin' or 'epoch' *li yuan* 曆元, the moment (usually in the relatively remote past) at which the heavenly bodies were presumed to have been in some standard starting configuration, such as a general conjunction at the winter solstice. This was the point from which all reckoning began. Han dynasty disputes on calendrical matters seem to have stressed disagreements about the choice of a system origin as much as disagreements about the choice of fundamental constants.[105]

It is clear from a reading of the *Zhou bi* that it contains a good deal of material relevant to calendrical astronomy. Some sections give detailed and explicit treatments of particular points, while others reveal significant background assumptions. It is not however possible to read the text as a deliberate exposition of a particular calendrical system. In the first place one important point is missing: there is no discussion whatsoever of the question of system origin. This alone makes it impossible to run a calendar on the basis of the *Zhou bi*. In the second place, despite the internal coherence of certain sections there is no evident order in the treatment of calendrical matters within the work as a whole. The calendrical content of the sections of the *Zhou bi* are listed below:

#A: Nothing of any relevance.

#B: The lengths of noon shadows at the solstices are given. Rong Fang asks about the amount of the sun's daily motion, but this question is not answered in the text now before us.

#C: Nothing of relevance.

#D: We are told of the 24 equally spaced 'solar seasons' qi into which the interval between successive returns of the same solstice is divided. The sun is in lodge

[104] The general question of the unofficial character of the *Zhou bi* will be taken up in chapter 3.

[105] See for instance the complaint of Feng Guang 馮光 in AD 175 that the use of a system origin not sanctioned by scriptural warrant was leading to widespread social disorder, and the reply by Cai Yong 蔡邕 given in a great assembly of officials: *Hou Han shu, zhi* 2, 3037ff.

Well at the summer solstice, and in Ox at the winter solstice. The solar cycle has 365 1/4 days; so circles concentric with the pole may be divided into 365 1/4 *du*.

#E: Nothing of any relevance.

#F: An (apparently) imaginary circumpolar celestial body is to be observed at midnight on the summer and winter solstices, and at 6 p.m. and 6 a.m. at the winter solstice.[106]

#G: An impractical method is described for measuring the extents of the 28 lodges by gnomon sightings. A fudged calculation is made of the sun's north polar distance at the solstices and equinoxes.

#H: A table is given of noon shadows of an eight-chi gnomon for each of the 24 qi.

#I: Systematic calculations are given for the amount by which the moon will have shifted to the east of its former position after certain calendrically significant intervals.

#J: Schematic data are given for the sun's rising and setting directions at the solstices.

#K: The month, the day and the year are defined. The lengths of five calendrical cycles are given. Calculations are set out justifying the values used for the length of the year in days, the daily motion of the moon, the mean number of lunar months in a year, and the length of a mean lunation.

It is evident, therefore, that only sections #I and #K deal very directly with matters of calendrical calculation. If the text of the *Zhou bi* was, as I shall argue later, a result of a process of accretion, it may therefore be that the material of greatest calendrical relevance was added last of all. Since #K deals with more basic issues it will be discussed first.

The quarter-remainder system type: basic structures

In #K1 we start with the most fundamental calendrical definitions:

> #K1[75a]
> Therefore a conjunction of the moon with the sun makes one month. When the sun returns to [the previous position of] the sun [in the sky], that makes one day. When the sun returns to a star, that makes one year.

The next two paragraphs make brief reference to the systems of the *heng* 衡 and the *qi* 氣 explained elsewhere in the *Zhou bi*. Next, without explanation of their function, we are given a list of five calendrical cycles:

[106] The author appears to have forgotten that since the year has 365 1/4 days a given stellar configuration will only repeat at a given time of day on a given solar date once every four years. This reinforces other evidence that we are not dealing with an actual observation.

#K4[75e]

The reckoning of Yin and Yang, the methods for days and months:

19 years make a *zhang* 章 [the 'Rule Cycle'].

4 *zhang* make a *bu* 蔀 [the 'Obscuration Cycle'], 76 years.

20 *bu* make a *sui* 遂 [the 'Era Cycle'], and a *sui* is 1520 years.

3 *sui* make a *shou* 首 [the 'Epoch Cycle'], and a *shou* is 4560 years.

7 *shou* make a *ji* 極 [the 'Ultimate Cycle'], and a *ji* is 31 920 years. All the reckonings of generation come to an end, and the myriad creatures return to their origin. [From this new origin,] Heaven creates the chronological reckoning [once more].

The functions of four of these cycles within quarter-remainder systems have been outlined earlier (page 24ff.): note the slight differences of names used here. It may be recalled that the 19-year cycle is the shortest common multiple of the solar cycle of 365 1/4 days and the mean lunation, while the 4560-year cycle gives a complete return to system origin conditions for the year number in the 60-fold cycle, day number, time of day, conjunction of sun and moon, and *qi* inception. As mentioned in the earlier discussion, the 31 920-year cycle is unique to the *Zhou bi*, and its function is not completely clear.

The text now proceeds to show how the lengths of certain important periods may be determined. Paragraphs #K9 to #K11 indicate that the length of the 365 1/4 day year may in principle be derived by observing that the length of noon shadow near a given solstice only repeats exactly at intervals of

$$(3 \times 365 + 366) \text{ days} = 1461 \text{ days}.$$

There may appears to be some disorder in what follows. Paragraph #K12 effectively repeats results given in #I6 and #I1 for the displacements of the moon against the background of the stars in a year and in a day, and #K13 points out the implication that while the sun makes 76 circuits of the stars the moon will perform 1016 circuits. Paragraph #K14 gives an apparently pointless restatement of the moon's daily motion.

The last two paragraphs are however orderly enough. Paragraph #K15 determines the number of mean lunations in one solar cycle of 365 1/4 days by using the equivalence of 76 years and 940 lunations, obtaining the result of 12 7/19 months. To find the length of a mean lunation (#K16), we divide 365 1/4 by this figure to obtain 29 499/940 days.

While there is nothing surprising in these calculations, we may note the hierarchy in which the data are arranged. First of all comes the list of cycles, including that of 76 years, the shortest number of years which is also a whole number of days and of mean lunations. This is of course based on the shorter cycle of 19 solar cycles and 235 lunations which lies behind the system of intercalary months. As already argued this is likely to represent an observation made during early experience of keeping lunations and seasons in rough step over a fairly long period. From this datum we determine the number of mean lunations per year. We then bring in the solar cycle of 365 1/4 days, and hence deduce the number of days in a mean lunation. It will be evident how very far this result is from any attempt to time a lunation directly, a procedure that would necessarily have produced confusingly varying results due to the irregularity of the motion of the sun and moon, and to the problems in defining phase visibility at all precisely.

In structuring its reasoning this way, the *Zhou bi* is in fact following the usual thought pattern of Han astronomers. Before the listing of standard constants in its description of the quarter-remainder system promulgated in AD 85 the *Hou Han shu* has a short discussion which is paralleled by that in the *Zhou bi*. In part it runs:

> In the derivation of calendrical constants, one sets up instrumental gnomons to compare solar shadows. When the [noon] shadow is longest then the sun is most distant: this is the starting point of the degrees of heaven [i.e. the winter solstice]. The sun leaves its starting point, and makes a circuit in a year, but the [noon] shadow does not repeat. After four circuits, i.e. 1461 days, the shadow repeats its starting [value], and this is the conclusion of the [whole pattern of] the sun's movement. If you divide the days by the circuits, you get 365 1/4, which is the number of days in a year. The sun moves one *du* in a day, so this is also the number of *du* in [a circuit of] heaven. If you observe the sun and moon setting out together from the *du* at which they start [at system origin], the sun does 19 circuits and the moon does 254 circuits before they return to meet at the starting point, so this is the conclusion of [the whole pattern of] the moon's movement. If you divide the moon's circuits by the sun's circuits, you get the number of circuits [performed by the moon] in a year. If you subtract the one circuit performed by the sun [in that time], the surplus of 12 7/19 is the number of times the moon has performed a circuit past the sun, which is the number of months in the year. If you divide that into the number of days in a year, you get the number [of days in] a month. (*Hou Han shu, zhi* 3, 3057–8)

Section #I is systematic enough to be read with little commentary. It begins by calculating the daily motion of the moon relative to the stars from the equivalence 19 solar cycles = 235 lunations and goes on to make similar calculations for small years (no intercalation) large years (with intercalation) and mean years, and then for short months (29 days), long months (30 days) and mean months. The results so obtained might be useful in avoiding large numbers in any calculations of the moon's future positions from a known position. Once more the lunar data are found from the use of a long-term solar-lunar cycle.

The sun, stars and seasons

The *Zhou bi* contains a number of statements relating to the seasonal position of the sun relative to a stellar reference system. We have, in summary form:

#D5: Summer solstice sun in Well; winter solstice sun in Ox.
#G9,10 [by implication]: winter solstice sun in Ox.
#G11,12 [by implication]: equinoctial sun in Harvester and Horn.
#G13,14 [by implication]: summer solstice sun in Well.

I will discuss these references together, postponing for the moment the treatment of one more related piece of evidence that does not quite fit the same pattern. There is a tradition of using such data as a means of dating an astronomical text such as the *Zhou bi*. One identifies the asterisms referred to, and then uses one's knowledge of the effect of precession to find at what epoch the sun was in the stated positions. I do not propose to follow this line.

In the first place there are obvious difficulties in identifying precisely which stars are referred to. We must not forget that the system of 28 lodges was not fixed in stone, and there is strong evidence that the determinative star of a given lodge could be changed. In addition, it must not be forgotten that ancient astronomers had no direct means of observing the sun's position amongst the stars. It could only be inferred indirectly from other observations. Our data are therefore already at one unverifiable remove from astronomical reality. More importantly, even if such data were originally calculated as objectively and accurately as conditions permitted (and this is just an assumption), we cannot assume that a later text is not using traditional data rather than up to date determinations. We know that official astronomers were being castigated for doing just this in the first century AD, so there seems no reason why an unofficial text such as the *Zhou bi* should be assumed to have done any better.[107] Clearly over very long periods it would be astonishing not to find the effects of precession causing noticeable changes in the contents of astronomical texts. I suspect, however that such considerations will be of little help in deciding where in history a particular section of the *Zhou bi* is to be placed.[108]

[107] See Jia Kui's 賈逵 discussion in *Hou Han shu, zhi* 2, 3027. This topic is treated in detail in Maeyama (1975–1976).

[108] Despite my suspicions about analyses of this kind, the reader might like to know the results of the calculations performed by Nōda (1933), p. 46. Using plausible identifications of stars (two possibilities for Well), he found the following limits during which the stated results held in fact:
Winter solstice in Ox: early eleventhth to mid-fifth century BC
Spring equinox in Harvester: late eleventh to early second century BC
Summer solstice in Well: early sixth century BC to early sixteenth/mid-eighteenth century AD
Autumn equinox in Horn: early sixth century BC to mid-fourth century AD

We are on rather firmer ground when we turn away from the real heavens and look at what astronomers said about the heavens. A series of datable texts is available in which we are told the sun's main seasonal positions amongst the stars. It is worth while comparing these with the contents of the *Zhou bi* to see if any obvious conclusion presents itself.

We may turn first to the *Yue ling* 月令 'Monthly Ordinances' chapters of the *Lü shi chun qiu* 呂氏春秋 (239 BC), where we find the following data for seasonal solar positions:

chapter 11 'the month of midwinter' [winter solstice]: Dipper.
chapter 2 'the month of midspring' [spring equinox]: Straddler.
chapter 5 'the month of midsummer' [summer solstice]: Well.
chapter 8 'the month of midautumn' [autumn equinox]: Horn.

This agrees with the *Zhou bi* for summer and autumn. but for winter and spring the lodges one step to the west have been used. The problem in drawing any firm conclusions from such data is obvious from the fact the *Yue ling* refers to 'the month of midwinter' etc. Not only are we not told when in the month the sun is in the position stated, but the month itself is a lunar one and can therefore shift about thirty days relative to the solar cycle. We may wish to *assume* that what we are being given is a deliberate attempt to tell us a best estimate of the sun's winter solstice position (for instance) – but once again that is only an assumption.

Chapter 3 of *Huai nan zi* 淮南子 (*c*. 139 BC) has:

eleventh month: Ox.
second month: Straddler, Harvester.
fifth month: Well.
eighth month : Gullet.

The solstitial positions now agree, and one of the spring equinoctial lodges is also used in the *Zhou bi*, although the autumn sun has shifted one lodge to the east of that used in the *Zhou bi*. But the cautions given above still apply in view of the fact that the reference is explicitly to months rather than to specific points on the solar cycle.

Liu Xin's account of the *San tong li* 三統曆 in *Han shu* 21b, 1005–6, may be assumed to continue the basic structure of the *Tai chu li* 太初曆 of 104 BC:

Any temptation to conclude that this data set must originate in the resulting bracket 'early 6th to mid 5th century BC' will be more easily resisted in the knowledge that the official calendrical system in use between 104 BC and AD 85 used the same seasonal lodges as the *Zhou bi*: see below.

winter solstice: start of Ox.
spring equinox: 4th *du* of Harvester.
summer solstice: 31st *du* of Well.
autumn equinox: 10th *du* of Horn.

Here at last we have precise agreement with the *Zhou bi*. But lest we should feel that we are dealing with a simple case of an improving accuracy of observation, it is worth noting that the 'start of Ox' had not in fact been the true winter solstice position since the middle of the fifth century BC, since when precession had been shifting the solstice westwards at the rate of one *du* about every seventy years. That meant it had moved about five *du* back into the preceding lodge, Dipper, by the Grand Inception reform in 104 BC, and had shifted six *du* in all by the time Liu Xin wrote at the beginning of the first century AD. Clearly there was more (or perhaps less) to the choice of solar reference points than simple observation.[109]

But eventually the discrepancy was too large to be tolerated. In the new system promulgated in AD 85 we have (to the nearest *du*, see *Hou Han shu, zhi* 3, 3077–9):

winter solstice: 21st *du* of Dipper.
spring equinox: 14th *du* of Straddler.
summer solstice: 25th *du* of Well.
autumn equinox: 4th *du* of Horn.

Once again, however, a simple judgement will not suffice. We cannot simply assume that all over China everybody who thought about the heavens suddenly changed their minds overnight, so that the *Zhou bi* has to be dated before 85 AD. Otherwise what are we to make of the fact that in his commentary to the *Zhou li* Zheng Xuan 鄭玄 (AD 127–200) tells us again that the sun is in Ox at the winter solstice, and repeats the rest of the data given in the *Zhou bi*?[110] In any case, as will now appear, the *Zhou bi* is not even consistent with itself in such matters.

[109] For comparison, in the ancient Mediterranean world Neugebauer points out that in the first century AD the Babylonian position of the vernal equinox relative to the stars was still in use by astrologers, despite its being three centuries out of date: Neugebauer (1975) vol.2, 594–598. Astonishingly this datum was still being repeated in the medieval European literature as late as 1400. This sets Liu Xin's conservatism in a certain perspective.

[110] *Zhou li zhu shu* 26, 19a. It is interesting that Zheng claims that at the equinoxes the two half moons will occur in the solstitial lodges. Clearly this would have been a datum that could have been checked by the naked eye. This is perhaps a hint of how ancient Chinese astronomers may originally have solved the problem of how to locate the sun among the stars: one simply uses the half moon to check its positions three months away (and presumably the full moon to check the position six months away). Given the fact that the lunar phases will not in general fall at exactly the required instants, such a method would be inherently loose, and there would obviously be no hope of settling the point by calculation.

In paragraphs #K8 and #K13 the 'Establishment star' *jian xing* 建星 (π Sagittarii) is the point at which sun and moon are located at system origin. It may therefore be taken as the winter solstice position, since most systems took this as their starting point. In the third century AD Zhao Shuang tells us that:

> The six stars of [the asterism] Establishment are over Dipper. To say that the sun and moon start out from the Establishment star refers to [a situation] in the eleventh month when winter solstice and conjunction fall together. Practitioners of calendrical methods count the *du* starting from five *du* before [the beginning of] Ox, so the Establishment star is near [this point]. (68e)

Zhao's identification closely follows that of Sima Qian around 100 BC:

> [The asterism] Southern Dipper stands for the court. To its north is the Establishment Star. (*Shi ji* 27, 1310)

More interesting from the calendrical point of view is the statement by Jia Kui 賈逵 in AD 92:

> In the Grand Inception system, at the winter solstice the sun was at the beginning of [the lodge] Ox, [the position marked by] the central star of the [asterism] Ox.[111] The ancient [systems of] Huangdi 黃帝, Xia 夏, Yin 殷, Zhou 周 and Lu 魯 had the winter solstice sun at the Establishment Star, which is our modern Dipper star. (*Hou Han shu, zhi* 2, 3027)

Jia goes on to point out that for some time astronomers had been aware that the solstice was in fact falling five *du* short of the Grand Inception position, and was therefore in the 21st *du* of Dipper. It is important to note that confronted with this datum he does not suggest an hypothesis equivalent to the precession of the equinoxes. His approach rather is to point out that in addition to being in harmony with the five ancient systems mentioned, a winter solstice in Dipper also has the backing of scripture in the *Xing Jing* 星經 'Star Canon' of the legendary star–clerk Shi Shen 石申, and of the *Shang shu kao ling yao* 尚書考靈曜, one of the 'apocryphal books' which were much respected in the court of the Eastern Han. Supposedly ancient tradition and observation could therefore be claimed to be nicely in step. Jia's preference for what was in his view the most ancient source makes it plain that he was not thinking in terms of a steady secular variation in the position of the solstice equivalent to modern precessional theory.

[111] The distinction signalled here is an important one. One star of the asterism named Ox marked the beginning of the lodge associated with that asterism. Clearly, since the middle star of Ox was used some of the asterism did not lie in the lodge of that name.

We may now sum up briefly. Sections #D and #G of the *Zhou bi* give solar seasonal positions representing the astronomically obsolete system that was given official support from 104 BC to AD 85. Section #K uses a different system that is (depending on one's point of view) more ancient or more modern. Once again, the only firm conclusion that emerges is that the *Zhou bi* cannot be understood as a single unified book.

Instrumentation and observation in the *Zhou bi*

The gnomon

The early history of the gnomon as an astronomical instrument is not easy to trace, whether in the ancient West or in East Asia. Perhaps the earliest reliable reference point is provided by the two mulAPIN ('Plough Star') cuneiform texts from Babylonia, so named from their opening words. While the copies we have today are dated around 700 BC they are said to be based on older material. In addition to other astronomical data, these tablets contain a table of shadow lengths of a gnomon of unspecified length for various times of day at the solstices and equinoxes.[112] An account given by Herodotus (*c.* 480–425 BC) suggests that Greek astronomers may have learnt of the use of the gnomon from Mesopotamian sources.[113] The first Greek shadow tables appear to date from around the fourth to fifth centuries BC.[114]

Firm early data are harder to find in China. There are references to the measurements of seasonal shadow lengths in the *Zhou li* 周禮 'Ritual of the Zhou [dynasty]', and although the length of the gnomon is unspecified we are given the value of one *chi* five *cun* for the summer solstice shadow.[115] Traditional Chinese scholars believed this book to have been compiled by the Duke of Zhou 周公 around 1000 BC; modern scholars would place it somewhere towards the end of the first millennium BC, probably before or near the beginning of the Western Han. Needham quotes an entry from the *Zuo zhuan* 左傳 chronicle for 654 BC, but although the Duke of Lu 魯 is said to have ascended to a high place 'to view the clouds and vapours' the references to shadow measurements supplied by Needham have no basis in the text apart from the fact that the record is dated to the winter solstice.[116]

112 Neugebauer (1975), vol.1, 544.

113 Dicks (1970), 166–7.

114 Neugebauer (1975) vol. 2, 740.

115 *Zhou li zhu shu* 10, 10a–11b. The Western Han commentators suggest that the gnomon was eight *chi* long, as in the *Zhou bi*.

116 Needham (1959), 284. The *Zuo zhuan* record comes from the fifth year of Duke Xi of Lu.

Some have tried to link the use of the gnomon for determining the dates of solstices to the adoption of the year of 365 1/4 days. Behind this suggestion lies the assumption that a good estimate of the length of the solar cycle cannot be obtained unless one observes two successive solstices accurately. In fact fairly rough estimates of solstice dates separated by a large enough number of years can also give accurate results. Further, one does not need to observe solstices at all. Other more easily observable astronomical events repeat with the same period, such as the dusk or dawn culmination of a given star. It is such observations rather than the use of the gnomon that are mentioned in the *Yao dian* 堯典, perhaps the earliest reference to the operation of the Chinese calendar.

The earliest firmly dated Chinese astronomical material referring to solstices is found in the 'Monthly Ordinances' chapters of the *Lü shi chun qiu* 呂氏春秋, which was completed in 239 BC. As with the other months, the months of midsummer, midwinter, midspring and midautumn are principally identified by the position of the sun among the lodges and the directly observable data of dusk and dawn centred stars. In addition, we have the following statements:[117]

Midspring: 'In this month [there occurs] the [equal] division of day and night.'
Midsummer: 'In this month, the longest day comes.'
Midautumn: 'In this month [there occurs] the [equal] division of day and night.'
Midwinter: 'In this month, the shortest day comes.'

There is clearly nothing here to force us to believe in the use of shadow measurements as basic seasonal data. But about a century later we are on firmer ground. In the book compiled for Prince Liu An around 139 BC we are told, amongst various data on the seasonal behaviour and growth of birds, animals, insects and plants:

> On the day of the winter solstice ... For an eight *chi* 尺 length (*xiu* 脩), the noon shadow is one *zhang* 丈 three *chi* 尺. On the day of the summer solstice ... the length of the shadow for eight *chi* is one *chi* five *cun* 寸 (*Huai nan zi* 3, 5b, *Sibu Congkan* edn)[118]

117 See *Lü shi chun qiu*, chapters. 2.1, 5.1, 8.1 and 11.1.

118 At the end of this chapter a short text has been appended which discusses gnomons in much more detail. It has been translated and discussed in Cullen (1976). Whereas the main text refers to its gnomon as 'a length' *xiu* 修, the additional material uses what is later the standard term *biao* 表 found in most of the *Zhou bi*. The gnomon is however ten *chi* high, a length found nowhere else. This material seems to have been added to the *Huai nan zi* text at some point. Whether it was composed later or earlier than 120 BC we cannot now tell. A.C. Graham has advanced the plausible speculation that this section, and possibly even the *Zhou bi* itself, may have originated within the Mohist school of the late Warring States period. See his *Later Mohist Logic, Ethics and Science*, 369–371.

Thirty-five years after the date of this work we are told that, as part of the work preparatory to the Grand Inception reform of 104 BC, Sima Qian and his colleagues *li gui yi* 立晷儀 'set up shadow instruments'.[119] It seems therefore that we can be sure of the astronomical use of gnomon shadows for the early Western Han, even if the picture is vague for earlier periods.

The *Zhou bi* uses three different terms for the gnomon. Of those sections that mention the gnomon explicitly, #F and #G use only the word *biao* 表.[120] In section #B, in which Chen Zi introduces the whole topic, the first reference to a gnomon is simply as a 'post' *gan* 竿 (#B9). In #B10 to #B12 we find *zhou bi* 周髀 or just *bi* 髀, then *biao* 表 in #B14. In #B15 Rong Fang, perhaps understandably confused, asks what the *zhou bi* is, and is told that it means a gnomon *biao* as anciently used at the capital of the Zhou kings.

The standard dictionary meaning of *bi* 髀 is 'thigh' or 'thigh-bone'. Why such a word should have been chosen to describe an upright pole about the height of a man is a little difficult to see. Outside the *Zhou bi* there appear to be no independent uses of the word in this specialised sense. In #B10 Chen Zi says that while the shadow represents the base, the *bi* gnomon is the altitude, the word for which, *gu* 股, also means 'thigh'. Any upright is, it seems in some sense a 'leg'. It also appears likely that there is a link between *bi* and the common word *bei* 碑'stele, pillar', which has the same phonetic but is written with the 'stone' radical rather than with 'bone'. The Eastern Han commentator on the *Yi li* 儀禮[121] discusses the customary placement of a *bei* stele within the complex of gate, courtyard and hall, and tells us:

> A palace must have a *bei* in order to observe the solar shadow (*Yi li zhu shu* 21, 14b)

Such a link would certainly make the use of *bi* for a gnomon rather less puzzling.

Observations of noon shadows

The Babylonian and early Greek shadow tables mentioned above gave values for various hours of the day, but in accordance with its emphasis on meridian observations all the solar shadows mentioned in the *Zhou bi* are noon values. Presumably one would ascertain the moment of noon by observing when the shadow crossed the meridian line through the gnomon; a simple method for defining such a line is given in #F7. Since the length of the shadow changes most slowly around noon, errors in

[119] *Han shu* 21a, 975. It must be said that the reference to instruments is missing from the description by Sima Qian himself.

[120] #G4 does however mention the use of a 'displacement marker' *you yi* 游儀 as a sighting aid on the circumference of its graduated circle marked on the ground. This sounds like another gnomon.

[121] A compendium of ritual practice which may be as late as the Western Han.

alignment are not critical. In addition the only values derived from observation seem to be solstitial values, and it is near the solstices that noon shadow-length changes most slowly from day to day. But as we shall see it would be a mistake to treat the observations in the *Zhou bi* simply as raw data suitable for objective scientific analysis. They are in fact significantly shaped by theory – a phenomenon not unknown elsewhere in the history of science.[122]

Solstitial shadows The principal shadow data in the *Zhou bi* are its values for the solstitial noon shadows of an eight *chi* gnomon. These are:

> Summer solstice: 1 *chi* 尺 6 *cun* 寸 = 1.6 *chi* (#B10,#D18,#H2).
> Winter solstice: 1 *zhang* 丈 3 *chi* 5 *cun* = 13.5 *chi* (#D18,#H2).

It is however impossible to understand the nature of these data without taking them together with the fact that the artificial 'shadow' resulting from a determination of the position of the north celestial pole is said to be 1 *zhang* 3 *cun* = 10.3 *chi*. Together with the application of the shadow rule (see below) the implication of these three quantities is that the radius of the sun's daily path round the pole at the winter solstice is exactly twice that at the summer solstice: see #B16 and #B18. This is clearly not simply a lucky coincidence. The observations have been influenced by the possibility of obtaining a theoretically neat result. Observation and theory have certainly both played a role, and we are not in a position to tell where one ends and the other begins.[123]

With this caution in mind, it is worth while analysing these data trigonometrically, if only to see how far theory has moved them away from any practical likelihood. Let us assume for the sake of argument that the two solstitial shadows belong together. Figure ar shows how the shadow lengths are related to latitude and to the obliquity of the ecliptic, which is the angle between the ecliptic and equatorial planes. It is easy to show that with the values of the *Zhou bi*:

[122] On Ptolemy's shaping of his numerical data, see Toomer (1984), viii, and also more controversially Newton (1977). In an age when accurately observed data were few and far between, the decision to give theory precedence over the messy particularities of one's data made good sense.

[123] We have other evidence of an ancient Chinese interest in concentric circles with diameters in simple ratios in the idealised specifications for ritual jade discs given in the *Er Ya* 爾雅 see *Er Ya zhu shu* 5, 18a. The ratios given there relate the diameter of the central hole, *hao* 好 to the thickness of the jade annulus which surrounds it, *rou*. 肉 If *hao*:*rou* :: 1:2 we have a *bi* 璧. If the ratio is 2:1 we have a *yuan* 瑗 and if the ration is 1:1 we have a *huan* 環.

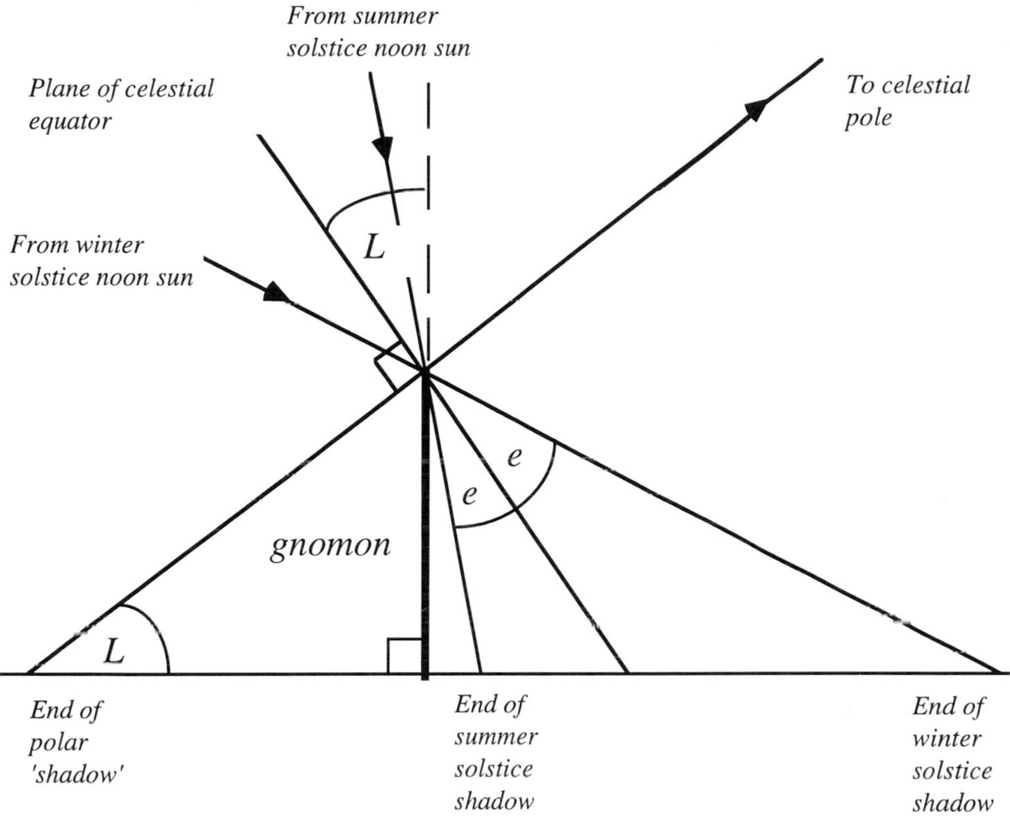

Figure10. Shadows, latitude (L) and obliquity of ecliptic (e).

Latitude $L = 35.33^{\circ}$ N
Obliquity $e = 24.02^{\circ}$.

However, if we are to treat the *Zhou bi* shadow lengths as observed values they must necessarily be subject to experimental error. An inch of error in the winter solstice shadow will alter both latitude and obliquity by about 6', whereas the same change in the summer solstice shadow changes the angles by 20'. It therefore seems more realistic to limit ourselves to saying that if the shadows are taken to be a consistent pair originating from observation they suggest a latitude of about 35° and an obliquity of about 24°. The latitude is consistent with a location near the ancient centres of Chinese culture in the Yellow River basin, while the slow change in the value of obliquity gives us no reason to rule out any date within the lifespan of Chinese culture up to the present day.[124]

The polar 'shadow' of 10.3 *chi* does not involve the obliquity: see figure ar. It follows directly from this datum that:

Latitude = 37.84°, or 38° in round numbers.

If we were dealing with real observations, the implication would be that the polar observation was made about 300 km to the north of the solar observations, which is rather far away from any likely locations for an ancient Chinese observer, since it could easily put him north of the Great Wall. Putting it another way, a polar 'shadow' at latitude 35° should have been about 11.4 *chi* rather than 10.3 *chi*.

In conclusion, then, we may say that it is possible that the solstitial shadow values may stem from an actual pair of observations. This impression is strengthened by the fact that these values are not repeated in other texts.[125] If this is so, the polar 'shadow' must be an artefact of the desire for a neatly arranged universe in which the summer and winter solar paths were related in size by a factor of two. Given the fact that the celestial pole cannot in any case be observed directly (see #F2 to #F4 for an indirect method) the degree of self-deception required would not have been very great.

The Table of Seasonal Shadows Section #H of the *Zhou bi* is devoted to a table of noon shadows of the standard eight-*chi* gnomon for each of the 24 *qi* 氣. As I shall shortly explain, it is fairly clear that the text before us is not the original, but is a

[124] Compare the discussion in Needham (1959), 286–91, particularly fig.113 showing the variation of obliquity with time. It is enlightening to compare the plausibility of the *Zhou bi* data with ancient Western parallels. Thus for example around 135 BC Hipparchus gave a ratio of 4:3 for gnomon length to equinoctial noon shadow at Athens. This yields latitude 37° rather than the correct 38°. Neugebauer comments 'Again we see the ration 4:3 is only a convenient estimate and is not the result of careful observation. Such primitive data were apparently never checked: at least Vitruvius, some 150 years later, still quotes the same ratio': (1975) vol. 2, 746. When in the first century AD Pliny gives similar shadow data for Alexandria, Babylon, Athens and Rome, it is only in the last case that his implied latitude error is less than half a degree – of course we would expect him to be less critical on such points than Ptolemy. Babylon is nearly two degrees out, Alexandria two and a half degrees, and Athens two thirds of a degree. One degree of latitude corresponds to about 110 km on the surface of the earth: Neugebauer (1975) vol. 2, 747–8. It is interesting that this is about the same as the figures given by several ancient Western authors who attempted to estimate the range within which gnomon measurements would not change perceptibly: see (1975) vol. 2, p. 726 n. 14. I am not trying to argue that ancient China was better or worse in these respects than the ancient West. The point is rather that the Western data give us some basis for judging how wrong purported observations can be before we have to suspect that they are completely fictional or deliberately mendacious. By this test the *Zhou bi* solstitial shadows look fairly realistic specimens.

[125] However, as pointed out in Cullen (1976) p. 124, the summer solstice shadow of a ten-*chi* gnomon in the anomalous *Huai nan zi* passage is two *chi*, corresponding precisely to 1.6 *chi* for an eight-*chi* gnomon.

reconstruction by Zhao Shuang. Before we turn to this question and to the shadow values themselves, there is something to say on the general topic of the *qi* (sometimes rendered as 'solar terms' or 'solar season'), the names of which are listed above each shadow length. The first time we meet the *qi* in a context of systematic mathematical astronomy is in the list given in Liu Xin's account of the *San tong li* 三統曆 in the *Han shu* (21b, 1005–6). For Liu Xin the *qi* begin at equal intervals throughout the solar cycle, which in his case is 365 385/1539 days long,[126] so that the interval between *qi* inceptions was 15 1010/4617 days (*Han shu* 21b, 1001). In systems of the quarter-remainder type as exemplified in the *Zhou bi* the interval is 15 7/32 days. Although the name of a *qi* can refer to the whole of the period until the next *qi* begins, in an astronomical context it is more usual to find that the instant of *qi* inception is intended.[127]

If we begin numbering the *qi* with winter solstice as #1, the odd-numbered *qi* are customarily designated 'medial *qi*', *zhong qi* 中氣, while the even numbered are 'nodal *qi*', *jie qi* 節氣.[128] The purpose behind the system as used by Liu Xin is to provide quick warning of the need for intercalation when the sequence of lunar months begins to slip out of step with the seasons. This is done by allocating one medial *qi* to each month, so that ideally *qi* #1, winter solstice, should fall at the inception of the first month in the series, and so on. In practice, all is well so long as each medial *qi* is still somewhere within its allotted month. Since the interval between two medial *qi* is over thirty days, the situation will eventually arise when a month contains no medial *qi*. This is then designated as an intercalary month, and the new month number is postponed until the next month (*Han shu* 21b, 1001).

A major function of the *qi* in the *San tong li* 三統曆 is therefore their ability to trigger intercalations at any time of the year, rather than having to wait until the year end to insert a thirteenth month, as appears to have been the practice in earlier systems. The fact that mid-year intercalations are found throughout the Western Han after 104 BC justifies us in thinking that the *qi* system was part of the basic equipment taken over from the Grand Inception scheme.

Further back than 104 BC, evidence for the *qi* as a complete system fades out quite quickly. Around 139 BC all twenty-four are mentioned in *Huai nan zi* chapter 3,

[126] It will be recalled that this apparent precision is due to the decision to combine a mean lunation of 29 49/81 days with the equivalence 19 solar cycles = 235 mean lunations.

[127] As Professor Nathan Sivin has pointed out (private communication) the principle behind the *qi* system is reminiscent of the '*tithi*' system of late Babylonian astronomy, in which instead of using real lunar months (each consisting of a variable whole number of days) one used for reckoning purposes a mean lunar month divided into 30 equal periods, each of which was close to a day long. See Neugebauer (1975), vol. 3, 1069.

[128] In the system of the seven *heng* circles described in section #D, the sun is exactly on a *heng* at the inception of a medial *qi*, and half-way between two *heng* at the inception of a nodal *qi*.

where however they are spoken of as following one another at fifteen day intervals: one suspects that we are no longer dealing with an astronomically precise division of the solar cycle. The excavated Linyi almanac for 134 BC mentions only the solstices and the beginnings of spring and autumn.[129] Before the Qin, the 'Monthly Ordinances' *Yue ling* 月令 chapters of the *Lü shi chun qiu* 吕氏春秋 (239 BC) only single out the solstices, equinoxes and beginnings of the four seasons as worthy of special notice. Some of the phenomena referred to in the names of the other *qi* are mentioned in passing, but the sense of the year being divided into twenty-four periods is absent.

This brief survey is necessary because of certain slight problems relating to the names of the *qi* in the *Zhou bi* shadow table. In the first place there is the question of order. The *qi* names in the *Zhou bi* follow the modern order, which is also that found in the *Huai nan zi* 淮南子.[130] In Liu Xin's listing, however, the names of *qi* #5 and #6 are interchanged, as are #8 and #9. The new calendar of AD 85 has them as in the modern order. Further, the name of *qi* #6 is *jing zhi* 驚蟄 'Arousal of [hibernating] insects' in both *Huai nan zi* and in texts from Liu Xin onwards. In the *Zhou bi* however the term is *qi zhi* 啟蟄 'Emergence of [hibernating] insects'. The occurrence of this term in the first month data of the *Lü shi chun qiu* as well as in the fragmentary and archaic *Xia xiao zheng* 夏小正 suggest that it might have been used in some listing of the *qi* older than any now extant. There would have been good reason to have altered the term by the time the *Huai nan zi* book was assembled under the patronage of Prince Liu An 劉安, for Qi 啟 was the personal name of his father Emperor Jing 景, who came to the throne in 156 BC. Use of the term for the rest of the Han was therefore taboo, and a replacement of similar meaning would have been used.

If therefore we assume that Zhao Shuang passed on the list of *qi* names to us unchanged,[131] what follows? The least adventurous deduction is that whoever wrote the original shadow table made no effort to follow the official practice of his day. If however we assume that the author followed whatever the orthodox fashion was, and that Liu Xin's practice was also that of the Grand Inception system, the order of the *qi* means that the original *Zhou bi* shadow table must date from after AD 85 or before 104 BC. Taken in conjunction with the archaic name, we would have to go back further than the accession of Emperor Jing in 156 BC, so that the old table would represent the earliest complete listing of the *qi*.

While the fact of the old table's crudity, not least in its use of a simple fifteen-day *qi* interval, would be consistent with such an early date, the whole argument rests on

129 See Chen Jiujin and Chen Meidong (1989).

130 I am not convinced by Qian Baocong's argument in Qian (1924) that the *Huai nan zi* originally had the same order as Liu Xin but was later altered.

131 This seems a reasonable assumption, since although he tells us he has altered the shadow lengths he says nothing about the names.

the assumption that all Han writers on astronomical topics thought and wrote in lockstep. How such unanimity could have been enforced I cannot think. Assuming it to have been enforced, how could one explain the innovations that did in fact occur? In any case it is well known that disagreement on calendrical matters was commonplace, and there is no reason to assume that the writers of the *Zhou bi* were exceptions to the rule. Looking at things from this perspective, we are faced with something of a non-problem.

Let us turn now to the shadow table itself. It contains identical solstitial shadow values to those already given in other parts of the text, and a glance reveals that it is constructed by simple linear interpolation between the solstitial values. Thus the two equinoctial values are both 7.55 *chi*, the arithmetic mean of the two solstitial values. Observation would have given a value about two *chi* less than this.

As I have already mentioned, the situation is complicated by the fact that it seems the present section #H was written by Zhao Shuang himself in replacement for the original *Zhou bi* table. The old table also based itself on linear interpolation between the solstitial values, but did so somewhat clumsily so that the equinoctial values came out slightly unequal. The table is reconstructed from Zhao's description in appendix 4. As Zhao points out, the old table is inconsistent with the description of the seven *heng* circles in section #D, according to which the distance of the sun from the pole (and hence from the observer at Zhou), should be the same at both equinoxes. It is likewise inconsistent with Chen Zi's statement in #B19. Once again the *Zhou bi* turns out to be a collection of documents disparate in detail if related in general theme.

We have already had occasion to note the strong resemblances between the procedures of the *Zhou bi* and the astronomical practice of the Western Han. The same resemblance can also be shown in the matter of shadow data. The basic pattern of linear interpolation between solstitial values was evidently something that everybody seems to have taken for granted up to the first century AD. Thus in the *Han shu* itself (26, 1294) we find the following set of shadow values:

summer solstice: 1.58 *chi*.
winter solstice: 13.14 *chi*.
equinoxes: 7.36 *chi*, the arithmetic mean of solstice values.

A series of prognostications then follows, based on advance or delay in the occurrence of predicted shadow lengths. Although this is not one of the sections of the *Han shu* that claims specifically to be summarising the writings of Liu Xin, creator of the Triple Concordance system, there is a strong chance that what we have here are Liu Xin's own values. Shadow lengths are not mentioned in the calendrical material of the *Han shu*, which quotes Liu Xin, but we do know that his father Liu Xiang (79–8

BC) used identical values in his *Hong fan wu xing zhuan* 洪範五行傳, for he is criticised for doing so in Li Chunfeng's commentary on the *Zhou bi*: see 30o–30q in Qian's edition. Li also quotes from the apocryphal work *Yi wei tong gua yan* 易緯通卦驗 (30o), and although he only gives solstitial shadows there is a much fuller quotation from what seems to be the same book in the commentary on the *Hou Han shu, zhi* 3, 3079–80. It has the same solstitial values of 1.48 *chi* and 13 *chi* and has both equinoctial values as 7.24 *chi*, clearly the result of linear interpolation. The work of Dull (1966) makes it seem highly likely that texts such as this were produced near the end of the first century BC and in the first decades of the first century AD.

In a Western Han context, there is it seems nothing at all unusual in the *Zhou bi*'s method of dealing with seasonal shadows. The first example of a Chinese attempt to give realistic values for shadows other than those at the solstices appears in the calendrical system of AD 85. *Hou Han shu, zhi* 3, 3077–9 gives a detailed table for the twenty-four *qi*, including the sun's north polar distance – clearly we are now in the context of *hun tian* practice. The values given include:

summer solstice: 1.5 *chi.*
winter solstice: 13.0 *chi.*
spring equinox: 5.25 *chi.*
autumn equinox: 5.50 *chi.*

As Maeyama has made clear nothing so simple as an outburst of open-minded empiricism and accurate observation has occurred here. The solstitial values are those given scriptural warrant by the *Zhou li* 周禮 and its commentators, as well as by at least one of the apocryphal scriptures so influential in calendrical matters in the first century AD.[132] It is also interesting to note that just as the original *Zhou bi* shadow table was not quite consistent with the *gai tian* scheme it seems from the inequality of the equinoctial shadows that the system of AD 85 could not quite match its numerical data to its new cosmography.

There is therefore nothing very odd about the *Zhou bi* shadow table data in a Chinese context . Evidence from the ancient West gives revealing parallels. One early Greek table (fourth or fifth centuries BC) gives seasonal shadow lengths for each of the twelve signs of the zodiac, using linear interpolation between solstitial values of 8 feet and 2 feet. This yields equinoctial values of 5 feet.[133] As Neugebauer comments:

[132] See the *Shang shu kao ling yao* 尚書考靈曜, quoted by Li Chunfeng in his *Zhou bi* commentary, 30f. in Qian's edition.

[133] The height of the gnomon is unstated: six feet would do for Athens, but a later text uses seven feet, which would suggest Babylon or Rhodes for the values given. The table also gives shadow lengths (again produced by a simple arithmetical scheme) for the hours of the day other than noon, the purpose evidently being to provide a simple method of telling the time. See Neugebauer (1975), vol. 2, 737–9.

> the exact trigonometric or graphical solutions of gnomon problems were preceded by a much more primitive phase in which simple arithmetical patterns provided reasonably close approximations of the actual variations of shadow lengths during the year and during the day. (Neugebauer (1975) vol. 2, 736)

The surprising thing about Europe is that, despite the great advantage possessed by Western astronomers after the Greek development of trigonometry, the use of such simple shadow tables continued to be popular long into the Middle Ages, although in progressively distorted form as generations of ecclesiastical copyists did their work.[134]

The shadow rule

We have already met the 'one *cun* for a thousand *li*' shadow rule by means of which several sections of the *Zhou bi* link measurements made at the gnomon to distances on a cosmic scale. The first piece of astronomical information given by Chen Zi to Rong Fang is:

> #B9 [26a]
>
> '16 000 *li* to the south at the summer solstice, and 135 000 *li* to the south at the winter solstice, if one sets up a post (*gan* 竿) at noon it casts no shadow. This single [fact is the basis of] the numbers of the Way of Heaven.'

> #B10 [26c]
>
> 'The *zhou bi* is eight *chi* in length. On the day of the summer solstice its [noon] shadow is one *chi* and six *cun*. The *bi* is the altitude [of the right-angled triangle], and the exact [noon] shadow is the base. 1000 *li* due south the base is one *chi* and five *cun*, and 1000 *li* due north the base is one *chi* and seven *cun*. The further south the sun is, the longer the shadow.'

This information is then applied in #B11 to find the height of heaven and the diameter of the sun. It is not until after all this that we find the explicit statement of the rule, in what seems to be a note interpolated by a later hand:

> #B12 [34h]
>
> Method: the *zhou bi* is eight *chi* long, and the decrease or increase of the base is one *cun* for a thousand *li*.

Shortly afterwards it is made plain that the rule can be used for finding the distance to the place below any point on the heavens, even if the sun is not there and the 'shadow' must be constructed artificially:

134 See Neugebauer (1975) vol. 2, 740–6. As Neugebauer sadly notes 'the western [i.e. Latin western European] tradition represents the lowest level in our material'.

> #B14 [34j]
>
> Now set up a gnomon (*biao* 表) eight *chi* tall, and sight on the pole: the base is one *zhang* three *cun*. Thus it can be seen that the subpolar point is 103 000 *li* to the north of Zhou.

Later on section #F describes sighting methods suitable for finding the 'base' in the case of a polar sighting. The shadow rule lies behind the calculation of the diameter of the *heng* circles 衡 in section #D, and is stated explicitly in #D18. It is used once more in #F, and is the ultimate basis of the data used in #G. It plays no role in the other sections of the *Zhou bi.*

It is not easy to trace the history of the shadow rule independently of the *Zhou bi* before the Eastern Han. The first datable mention of the shadow rule comes in Zhang Heng's 張衡 essay on the cosmos, *Ling xian* 靈憲, written about AD 100, in which he is describing the celestial sphere of the *hun tian* 渾天 universe. After telling us that the height of (presumably the zenith) of heaven above the flat earth is half of the 232 300 *li* diameter of the universe he continues:

> To cover these numbers, use repeated [applications] of base and altitude; the shadow coming down from the heavens and its significance on the lowly earth [are related by the fact that] every shift of a thousand *li* gives a [shadow] difference of one *cun.*[135] (*Hou Han shu, zhi* 10, 3216)

Somewhat later we have the evidence of the commentary on the *Zhou li* 周禮 by Zheng Xuan 鄭玄 (AD 127–200). He is explaining the section of the text describing the activities of the official bearing the title Da Si Tu 大司徒 'Grand Director over the Masses', in which there is a reference to the use of solar shadow observations to find 'the centre of the land' where the royal capital should be set up. In his comment Zheng states:

> In general there is a difference of one *cun* of solar shadow for one thousand *li* on the ground. (*Zhou li zhu shu* 10, 10a)

Once we try to go any further back than this the trail becomes very thin. At the end of chapter three of the *Huai nan zi* 淮南子 there is a section of material in which the 'one *cun* for a thousand *li'* rule is applied in conjunction with a gnomon ten *chi* high rather than the eight *chi* of the *Zhou bi*. As a consequence, heaven turns out to be 100 000 *li* above the earth rather than 80 000 *li*. It is however clear that this material is not part of the original text of *Huai nan zi* compiled around 139 BC, but was added by

135 The problem with this passage is that there seems to be no way of getting from plausible shadow measurements to the dimensions quoted by Zhang. On this problem, and on the later history of the shadow rule, see Fu (1988).

a later hand.[136] Of course the added material could still be quite old, and as I have already mentioned its form and content have been taken to suggest connections with the Later Mohist group in pre-Qin times. But all this is no more than supposition.

A final tantalisingly incomplete hint of an early use of the shadow rule may be found in the *Lü shi chun qiu.* As we saw earlier (page 52), this work gives a brief note of important features of a cosmography similar to that of the *Zhou bi.* Just before the passage quoted earlier, we are told that:

> Within the four poles, it is 597 000 *li* from east to west, and from north to south it is also 597 000 *li* (*Lü shi chun qiu* 13, 3b, *Sibu Congkan* edn)

The following text tells us that the sun 'goes round the four poles' (with 'poles' here in the sense of 'limits'), so the figure given looks as if it is the diameter of a solar path centred on the celestial pole. Now the largest solar path in the *Zhou bi* is the winter solstice path, which has diameter 476 000 *li.* This figure was obtained using the shadow rule in conjunction with an eight-*chi* gnomon, a winter solstice shadow of 13.5 *chi* and a 'polar shadow' of 10.3 *chi.* If the same rule and the same shadow data are used with a larger gnomon, then proportionately larger celestial distances result. If, for instance, we used the ten-*chi* gnomon found in the anomalous *Huai nan zi* text, the winter solstice solar path would turn out to have a diameter:

$$476\,000\ li \times 10/8 = 595\,000\ li.$$

This is within 2000 *li* of the figure given in the *Lü shi chun qiu.* One extra *cun* on either of the shadow lengths used would have made the agreement exact, since the diameter is double the radius. But of course all this is just conjecture.

Let us now sum up the historical situation. We have good evidence for the use of the shadow rule from the first century AD onwards. The *Huai nan zi* fragment presumably takes us some way back into the Western Han. Both it and the *Lü shi chun qiu* give us rather weak indications of an even earlier date. Clearly such evidence is not likely to be of crucial help in dating the relevant sections of the *Zhou bi.*

As to the origins of the shadow rule, I do not think there is any evidence on which to base worthwhile conclusions. The most striking fact about the rule is how completely wrong it is. Thus in #B10 Chen Zi told us that displacing the position of the summer solstice observation by 1000 *li* north or south would change the length of the shadow by one *cun.* In fact a simple trigonometrical calculation shows that a latitude change southwards of 0.68°, equivalent to about 75 km, would produce a reduction of summer

[136] See Cullen (1976).

solstice shadow from 1.6 *cun* to 1.5 *cun*. The distance involved is therefore something like 150 *li* rather than the 1000 *li* quoted.[137]

It would be rather easier to cope with this situation if we could find evidence that Chen Zi's 'one *cun* for a thousand *li*' was simply a hypothetical datum introduced for illustrative purposes. But there is no indication of this at all in the text. In the *Huai nan zi* fragment, which does use the hypothetical *jia shi* 假使 'suppose it to be the case that' in parts of its discussions, the shadow rule is also treated as a fact pure and simple, and the same is true of most of those who use the rule after the Han. Later in the first millennium AD doubts began to grow, and Li Chunfeng's commentary brings such empirical evidence as he can muster to show that the north–south movement for an inch of shadow cannot possibly exceed 500 *li*, and may well be less: see 30e to 31h in Qian's edition. He also points out that such a rule only makes sense if heaven and earth are flat and parallel planes. In the following century the great meridian survey of Yixing 一行 and Nangong Yue 南宮說 provided even clearer evidence that the old rule was faulty.[138]

It does seem however that in earlier times the rule was simply accepted on the basis of tradition. How that tradition originated we can only guess, though it can hardly have been on the basis of experimental evidence.[139] The fact that it survived as long as it did is highly revealing of the circumstances under which ancient Chinese astronomy developed. The only place where gnomon shadow measurements were likely to be made was at the official observatory at or near the capital. From the early Zhou until the end of the Han this was always located somewhere close to the parallel of 35° N latitude. No astronomer had any reason to leave his post to carry out observations elsewhere. Even if he or some private individual did so, it would have been extremely difficult for him to have translated a rough knowledge of the distance measured along winding roads into precise north–south displacement. The situation in the ancient Mediterranean world was a little more favourable in one sense, since astronomers worked at a number of geographically scattered locations in a culture much less centripetal than that of ancient China. But, as we have already seen, in the absence of

137 Of course the length of the *li* varied somewhat through the centuries, but never enough to account for the discrepancy encountered here.

138 See Beer et al. (1961). See also Needham (1959), 292, and (1962), 44, Cullen (1982) and Fu Daiwie (1988).

139 The only thing to be said in favour of the rule is that the notion of a linear variation of noon shadow length with north–south displacement does hold approximately over an appropriately limited range of latitudes. If for instance we calculate for the range 30° to 40°, we find that the change of summer solstice shadow for each degree of latitude remains within 3% of 1.45 *cun*/degree. But anyone who was aware of this linearity on experimental grounds would also have been aware that the 'one *cun* for a 1000 *li*' ratio was wide of the mark.

any systematic collection of data it was difficult for even latitudes to be accurately known, let alone actual distances over the surface of the earth.[140]

I cannot leave the topic of the use of the shadow rule without referring to a further aspect of the problem. Fu Daiwie (1988) has pointed out that a fascination with the vastness of cosmic dimensions can be traced back as far as the time of Zou Yan 鄒衍 in the fourth century BC. The theme can be followed through the *Lü shi chun qiu* in the third century BC, the *Huai nan zi* book in the second century BC, and the *chan wei* 讖緯 apocrypha around the time of Wang Mang. The important thing about the *Zhou bi* and related texts such as the fragment at the end of *Huai nan zi* chapter 3 is that they provide for the first time a simple and rational way of obtaining such dimensions from observation. Our focus should not be so much on the arbitrary nature of the shadow rule, but on its elegance and generality in comparison with the numerological luxuriance that lay elsewhere. From this point of view we can sympathise with Rong Fang's eagerness to learn the Way of Chen Zi.

Stellar observations with gnomons

Although we most naturally think of the gnomon as an instrument for measuring the shadow cast by the sun, we have already seen good evidence that it could be used in conjunction with a north–south sightline for observing the transits of celestial bodies at night. Independently of the *Zhou bi* we also have evidence that the gnomon could be used in other ways for stellar observations. Thus the 'Artificers' Record' *Kao gong ji* 考工記 section of the *Zhou li* 周禮 tells us of the activities of the master craftsmen in charge of laying out sites in the proper orientation:

> They check the ground with water levels. They set up a pole [and brace it] with cords to observe the shadow. Using compasses they check the shadows cast by the

[140] Even the data behind Eratosthenes' famous (and, it is usually claimed, accurate) determination of the earth's circumference are argued by one expert historian of ancient Western science to have been 'no more than crude estimates expressed in convenient round numbers': Neugebauer (1975) vol. 2, 653. The same is even more true of his successor Posidonius. For a more optimistic interpretation of Eratosthenes' work, relying heavily on the adoption of the right value for the stade, see Dreyer (1906), 174–7. Lloyd (1987), 231ff., reviews recent discussions on Eratosthenes. On the general question of the gathering of geographical data in Western antiquity, Neugebauer's pessimism certainly seems justified: 'In spite of complete theoretical insight into the astronomical requirements for the determination of accurate geographical coordinates the absence of any scientific organisation in antiquity reduced geography to the unsatisfactory game of constructing compromise solutions from accidental records of travellers and merchants, from the writings of historians, or, perhaps, some Roman itineraries and army records. Not before the time of Biruni (i.e. around AD 1000) was there a more systematic approach possible': (1975) vol. 2, 938. His point is greatly strengthened by the fact that despite a lack of the vital concept of the sphericity of the earth, it was in China three centuries before Biruni that the first great meridian survey occurred under the leadership of the monk mathematician Yixing, because the Tang dynasty astronomical bureaucracy was able to draw on the wealth and organising power of what was then the world's greatest and most culturally advanced empire.

rising and setting sun. By day they observe the sun's shadow [in this way], but by night they examine matters by [observation of] the pole star, in order to fix east and west (*Zhou li zhu shu* 41, 23a–24b.)

In some cases, as we shall see, the use of the gnomon for stellar observations in the *Zhou bi* turns out to be rather more problematic than the examples we find elsewhere.

Artificial shadows The first reference to a stellar sighting in the *Zhou bi* is an essential part of Chen Zi's discourse on the dimensions of the universe:

#B14 [34j]
Now set up a gnomon (*biao* 表) eight *chi* tall, and sight on the pole: the base is one *zhang* three *cun*. Thus it can be seen that the subpolar point is 103 000 *li* to the north of Zhou.

This seems straightforward enough at first glance. One simply sights over the top of the gnomon towards the celestial pole, and produces the line of sight back to the ground to mark the tip of the 'shadow' that would have been cast if one had been sighting on the sun. As we shall see, later sections describe the obvious use of a stretched piece of string for this purpose. The shadow rule may then be applied to find the distance from the observer of the point below the pole.

The problem is, however, that in Han times there was no bright star very close to the pole as there is today. It seems that Chen Zi's purported observation is very much a thought experiment. How then did one know where the celestial pole was? The principle emerges from section #F, which begins with the words 'If you want to know the pivot of the north pole ...', which neatly states our problem. The material that follows has two aspects. One is quite unproblematic: the text explains the use of the gnomon and string sighting device to observe a bright circumpolar star as it orbits the pole, and hence to locate the pole and identify the cardinal directions. The problem lurking in the text is that, as we shall see, the star referred to cannot be easily identified with an astronomically plausible object, and appears to have been placed as it is for reasons that can only be described as subterfuge.

For the moment, we will ignore the hidden problem. The method for observing the star in question is set out in paragraphs #F1 to #F5. In summary, one observes the star in four positions relative to the pole:

#F2: extreme east and west positions. This yields two 'shadows' whose ends are 2 *chi* 3 *cun* apart, revealing (by the shadow rule, though not explicitly stated) that the two positions of the star are 23 000 *li* apart, and defining and east west line between the shadow ends. The line from the gnomon to the mid-point of the line joining the shadow end points north–south.

#F5: extreme south and north positions (superior and inferior transits in modern terms). This yields two shadows on a north–south line, one 11.45 *chi* long and the other 9.15 *chi* long. These translate into distance as 114 500 *li* and 91 500 *li,* once again making the two extreme positions of the star 23 000 *li* apart, and implying that the pole itself is 103 000 *li* away, since it must be half-way between the extreme positions in accordance with the implied assumption of the shadow rule that heaven is a flat plane parallel to earth.

It would clearly be quite pointless to try to translate the situation described here into the language of celestial spherical trigonometry, since it is obvious that the mind that produced this text was thinking in terms of a flat heaven.

The Width of Lodges Other uses of the gnomon in the *Zhou bi* are decidedly problematic. We turn first to the early part of section #G. In this an attempt is made to use gnomon observations to define the widths of the 28 lodges. Now as we have seen it is highly likely that the combination of gnomon and waterclock was indeed the method by which precise lodge widths were originally defined. The point here, however, is that the text dispenses with the waterclock. The procedure begins with instructions to describe a circle with circumference exactly 365 1/4 *chi* on level ground.[141] The circumference thus gives us one *du* for every one *chi.* We then fix north–south and east–west diameters, and erect a gnomon at the centre.
The lodge chosen as an example in the text is Ox, where the sun is located at the winter solstice. Once the circle and gnomon are set up we proceed as follows:

> #G4 [58n]
> ... sight on the centring of the middle star of Ox. Next further observe the leading star of Woman. As previously, sight with the gnomon and string on the leading star of Woman, so as to fix its [moment of] centring. Thereupon, using a displacement marker, sight [again] on the middle star of Ox, [noting] how many *du* it is to the west of the central standard gnomon. In every such case, the number of *chi* indicated by the marker gives the number of *du*. The marker will be over the eight *chi* [mark], so we know that Ox has eight *du*. Proceed accordingly with succeeding stars, so that all the 28 lodges are determined.

The situation is straightforward. We watch the determinative star of the lodge Ox (which happens to be the central star of the asterism of that name) cross the meridian. We then wait until the determinative star of the next lodge to the east of Ox, Woman, crosses the meridian, and take another sight on the previous star. To do this we will

[141] This is supposed to be done by making the diameter of the circle exactly 121 *chi* 7 *cun* 5 *fen*, which assumes $\pi = 3$. This is of course the first sign that we are dealing with a procedure whose author has never carried it out in practice.

have to move our sightline so that we are now looking somewhat west of south, and we mark the point where our sightline crosses the graduated circle. At that point we read off the number of *du*, which (we are assured) gives us the extent of the lodge Ox.

In fact, however, this is impossible. What we are doing is measuring the difference in azimuth between the two celestial bodies. This is an angle measured in the plane of the horizon, whereas we require the Chinese equivalent of right ascension, which is (in modern terms) measured in the plane of the celestial equator. To see the effect of this mistake, let us take some examples.

The two stars referred to in the *Zhou bi* are easily identified, and it is simple to find their coordinates for the epoch 100 BC. This epoch is chosen to represent Western Han conditions. We have:

	RA (degrees)	Dec. (degrees)
Ox: β Cap	275.10	–18.80
Woman: ε Aqr	282.84	–14.84

The difference between right ascensions (RA) of these stars is thus 7.74° which converts to 7.85 *du*. Clearly the lodge width is as stated in the *Zhou bi* and in other Han texts. If however we follow the procedure of the text the result is different. When the determinative star of Woman is on the meridian we have:

	Azimuth (degrees)
Ox:	170.97
Woman:	180.00

The difference is thus 9.03° or 9.16 *du*, over one *du* greater than the 8 *du* promised in the text. In most other cases the discrepancy is even more marked. Thus the width of the lodge Mane, whose determinative star is η Tauri, comes out at 10.45° in right ascension, which fits in well with the ancient value of 11 *du*. The method prescribed by the text would have yielded no less than 28.26°, a glaringly obvious discrepancy. Even if the author of this section of the *Zhou bi* tried his method for the two stars he mentioned and convinced himself that the answer was 'about right', he could not have continued in his self-deception for very long.

In the universe of the *Zhou bi* it is not very difficult for us nowadays to see that azimuthal observations of the kind specified here will only yield true lodge widths if the observer is situated at the subpolar point, so that the radii of his graduated circle coincide in plan view with those radiating from the pole. As we have seen, however, the concept of angle was not well developed in ancient China, and what seems obvious to us with hindsight may not have seemed obvious to the author of this section, who may well have believed that his idea would work.

We cannot be sure what his purpose was in creating this new method for measuring lodge widths. What after all would have been wrong with transit-timing using a waterclock? The use of the waterclock is indeed mentioned in #F3, though not in the present context. I suspect that the evident wish of the writer to be able to read off celestial coordinates from a graduated circle suggests that he wanted to emulate the performance of *hun tian* 渾天 instruments. This suspicion becomes a certainty when we study the succeeding paragraphs of section #G, which we shall shortly do.

First, however, let us conclude the discussion of the graduated circle. Once the widths of all the lodges have been established, we are told that they must be marked out on the circumference of the circle: see #G6. As we have already seen, the manner in which this is to be done indicates strong links between this portion of the text and the *shi* 式 divination devices discussed earlier. In #G8 we are told to use the circle to determine in what lodge the sun is setting and rising. It is clear that in this case the lodge system is simply being used as an azimuth reference system for a terrestrial observer – and it is indeed marked out round the edge of the fixed earth-plate of the *shi* as well as on the rotating heaven-disc. There is of course nothing illegitimate in such a procedure,[142] although it seems to be more common to divide the horizon using the twelve cyclical signs, *zi* 子 to *hai* 亥. The eight trigrams, *ba gua* 八卦 are also sometimes used. Both of these types of graduation are found on the earth-plates of *shi*, and both are exemplified in the brief notes on rising and setting in section #J.

North polar distances Section #G concludes with three statements, each followed by 'method' paragraphs explaining the derivation of the results given. The key statements are:

> #G9 [60n]
> The distance of Ox from the pole is 115 *du* 1695 *li* 21 *bu* and 819/1461 *bu*.

> #G11 [62h]
> The distance of Harvester and Horn from the pole is 91 *du* 610 *li* 264 *bu* and 1296/1461 *bu*.

> #G13 [63b]
> The distance of Well from the pole is 66 *du* 1481 *li* 155 *bu* and 1245/1461 *bu*.

Let us for the moment leave to one side the (to us) odd mixture of circular and length units. We have here a set of north polar distances which although they ostensibly refer to the asterisms which lend their names to lodges are clearly intended to tell us

142 It seems possible that the legitimacy of using the lodges as terrestrial azimuth graduations may have led the writer of section #G into his confusion about the measurement of celestial lodge widths.

the north polar distances of the sun when it is passing through the lodges in question. Ox is the winter solstice lodge, Harvester and Horn are the equinoctial lodges, and Well is the summer solstice lodge. These data agree closely with those given in Jia Kui's 賈逵 memorial of AD 92:

> [it has been stated that] at the winter solstice the sun is 115 *du* from the pole, that at the summer solstice it is 67 *du* from the pole, and that at the spring and autumn equinoxes it is 91 *du* from the pole. (*Hou Han shu*, *zhi* 2, 3029)

The problem is, however, that within the context of the *Zhou bi* the mere appearance of north polar distances in *du* is the equivalent of the clock striking thirteen. There are two reasons why these data are so anomalous. In the first place, the methods of measurement described in the *Zhou bi* are quite incapable of observing such data. The simplest procedure for finding the north polar distance of the sun directly would require a graduated ring fixed vertically in the plane of the meridian. The sun's zenith distance as it crossed the meridian could thus be combined with a previous determination of the zenith distance of the pole to give the solar north polar distance. But in the *Zhou bi* we are still in the situation mentioned in the *Han shu*, as cited earlier:

> Distances from the pole are hard to ascertain, so one has to use the shadow. It is the shadow that enables one to know the sun's north and south [displacements]. (*Han shu* 21a, 1294)

It would certainly be possible to obtain angular data from shadow measurements by geometrical construction – perhaps even using the giant graduated circle on the ground mentioned earlier. But nobody in the *Zhou bi* or elsewhere ever mentions such a procedure, and in any case it is clear from the precision claimed that the data before us cannot result from any such procedure, and indeed cannot be the result of observation at all.

There is another more fundamental problem to confront, which is that the entire concept of a north polar distance in angular measure is foreign to the astronomical system of the *Zhou bi*. The *du* is essentially a fractional division of the circumference of a circle. As we have seen in section #D, this implies that on a particular circle the *du* will correspond to a particular length round the circumference. In the *hun tian* universe north polar distance is measured along a great circle of the celestial sphere though the pole and the body in question. There is therefore no problem about expressing such a measurement in *du*. In the *gai tian* universe, the distance of a body from the pole is measured along a straight line which is a radius of the celestial disc. It is therefore incoherent to use the *du* to express such a distance. How would one know what length of *du* to use? And yet here we have a *du* which is clearly of a definite length.

The source of these puzzling data is revealed in the 'method' paragraphs which follow each statement. For simplicity I will discuss the equinoctial data first.

#G12 [62k]
Method:
Set up the distance from the middle *heng* to the pivot of the north pole: 178 500 *li*, which makes the dividend.
Take as divisor the size of one *du* on the innermost *heng*.
The integral quotient gives the number of *du*.
The *li* and *bu* are found from the remainder.
Make the [final] remainder the numerator of a fraction whose denominator is the divisor.

The middle *heng*, the fourth of the seven described in section #D, is the daily circle of the sun round the pole at the equinoxes. Its radius has been found by halving the figure of 357 000 *li* given in #D11. The innermost and first *heng* is the summer solstice path, of diameter 238 000 *li*, divided into *du* of length 1954 *li* 247 *bu* and 933/1461 *bu*: see #D8. There seems to be no reason whatsoever for taking the length of a *du* on this *heng* as the unit for measuring the radius of another *heng*. A little further analysis will however make the point behind this choice plain.

We must recall that the seventh *heng*, the winter solstice path, is exactly double the radius of the summer solstice first *heng*, and that the equinoctial fourth *heng* is half-way between the two. The radii of the first and fourth *heng* are thus related by:

$$R_1 = (2/3) \times R_{4.}$$

Taking π as 3, as throughout the *Zhou bi*, the length of a *du* on the innermost *heng* is:

$$d = (2 \times 3 \times R_1) \div (365\ 1/4)$$

so substituting from the first equation, we find:

$$\begin{aligned} d &= 2 \times 3 \times (2/3) \times R_4 \,/\, (365\ 1/4) \\ &= 4 \times R_4 \div (365\ 1/4). \end{aligned}$$

This is the unit we now have to divide into the radius of the middle *heng* as prescribed in the method given, in order to find the number of 'standard *du*' (as reckoned round the circumference of the innermost *heng*) equivalent to the radius of the fourth (equinoctial) *heng*. The result will be equivalent to the number of *du* in the north polar distance of the equinoctial sun:

$$R_4 \div [4 \times R_4/(365\ 1/4)]$$
$$= (365\ 1/4) \div 4$$

which is of course (in our terms) a quadrant or 90°, exactly the north polar distance required for the equinoxes. This result has nothing at all to do with observation. It follows straightforwardly from a decision by the author of this section to exploit the simple numerical relationships between the *heng* radii, which are themselves in turn the result of a decision to prefer neat theory over observation. The only reason the innermost *heng* has been used to provide the length of a *du* is because it produces the result required. The approximation $\pi = 3$ is also crucial.

Supposing that we allow that this is a legitimate way to convert *heng* radii to north polar distances. Since the radii of the summer solstice, equinoctial and winter solstice *heng* are related by:

$$R_1 : R_4 : R_7 :: 2/3 : 1 : 4/3.$$

This would yield north polar distances as follows:

summer solstice: $(2/3) \times (365\ 1/4) \div 4$ *du* = 60 *du* 7/8 or 60°
winter solstice: $(4/3) \times (365\ 1/4) \div 4$ *du* = 121 *du* 3/4 or 120°.

These are clearly not the results stated in #G9 and #G13. How then do the processes of calculation differ from that used for the equinoxes? The corresponding 'method' paragraphs are:

#G10 [60p]
Method:
Set up the distance from the outermost *heng* to the pivot of the north pole: 238 000 *li*.
Subtract 11 500 *li* for the *xuan ji* 璿璣, leaving 226 500 *li,* which makes the dividend.
Take as divisor the size of one *du* on the innermost heng: 1954 *li* 247 *bu* and 933/1461 *bu.*
The integral quotient gives the number of *du.*
The *li* and *bu* are found from the remainder.
Take 1/300 of the rationalised remainder as the dividend, and take 1461 as the divisor, which gives the number of *li.*
Multiply the remainder by three, and division yields hundreds of *bu.*
Multiply the remainder by ten, and division yields tens of *bu.*

Multiply the remainder by ten again, and division yields units of *bu.*
Make the remainder the numerator of a fraction whose denominator is the divisor.
Proceed similarly in subsequent cases.

#G14 [63d]
Method:
Set up the distance from the innermost *heng* to the pivot of the north pole: 119 000 *li.*
Add 11 500 *li* for the *xuan ji,* giving 130 500 *li,* which makes the dividend.
Take as divisor the size of one *du* on the innermost heng.
The integral quotient gives the number of *du.*
The *li* and *bu* are found from the remainder.
Make the [final] remainder the numerator of a fraction whose denominator is the divisor.

We can see at once what has happened. Before carrying out the conversion to *du* as before, 11 500 *li* has been subtracted in the case of the winter solstice and added in the case of the summer solstice. This is, as the text states, the radius of the circle described by the *xuan ji* 璿璣 'star' as it orbits the pole in accordance with the description in section #F. What is more, we may note that according to #F1 this star is on the same side of the pole as the winter solstice sun, but is on the other side from the summer solstice sun. It is therefore obvious why we are told to subtract 11 500 *li* for winter and add it for summer: the 'distances from the pole' we are calculating for the solstices are in fact measured from the *xuan ji* star and not from the pole at all. The equinoctial distance is however measured from the pole.

Now, as we shall see, it seems that the radius of the *xuan ji* circle is determined by other considerations than the need to produce convincing north polar distances for the solstitial sun. To us it may therefore seem that the author of this section was simply lucky that some simple manipulations of the data enabled him to produce results close to those accepted in the first century BC. It is however unlikely to have seemed that way to its creator, who was probably convinced that he had stumbled upon a highly significant fact about the cosmos.[143]

What can we deduce from this strange state of affairs? The main point is that no-one could possibly have created the peculiar computational procedures we have seen here unless he knew in advance what answer he wanted to get. Clearly, therefore, the author of this section already knew values for the seasonal north polar distances of

[143] A much more unlikely numerical coincidence seems to have been crucial in the creation of the universe of nested planetary spheres usually known as the 'Ptolemaic system': see Neugebauer (1975) vol. 2, 917ff. The background of numerical coincidences to Kepler's linking of the five regular solids to the orbits of the planets in the *Mysterium Cosmographicum* is too well known to need repeating here.

the sun. Since the methods of the *Zhou bi* cannot yield observational data of this type, the author must have worked at a time when *hun tian* astronomy was already being practised. As we have seen, that probably places him no further back than some way into the first century BC. It seems that he may have felt that it was essential for *Zhou bi* methods to do as well as the *hun tian*, a motive that as we have already seen probably lies behind section #F as well. To us, his success has the whiff of despair about it. At the time, however, an astronomical scheme that could deduce the values of important quantities from pre-existing data would probably have seemed superior to one that could only observe them.[144]

The problem of the xuan ji 璿璣 Before leaving the topic of stellar observations, we are now in a position to resolve various problems which arise from the description of observations of the *xuan ji* star in section #F. Two features require explanation.

(1) The star in question is said to be 11 500 *li* from the celestial pole.
(2) The star is placed on the radius linking the winter solstice sun to the pole.

The answer to the first question takes us back to the description of polar conditions set out by Chen Zi in #B22 and #B24, in which we are told that the rays of the sun just reach the subpolar point at the equinoxes, so that there is a six-month day and a six-month night there. This accurate statement of polar conditions implies that the range of the sun's rays should be equal to the radius of the equinoctial orbit of the sun round the pole (the fourth *heng*) which is 178 500 *li*. We are however told in #B25 that the range of the sun's rays, which is the same as the extent of human vision, is in fact only 167 000 *li*. While the text goes on to calculate how much the summer solstice sunlight at noon and midnight overlap across the pole, and what the corresponding shortfall is in winter, the now embarrassing calculation for the equinoxes is not made. So why was the appropriate value for the solar range not used?

The answer becomes clear in #B32 – see also #D15. By adding twice the range of the sun's rays to the diameter of the winter solstice orbit, one may define the circle of 'the four poles' which is the outer limit of the region ever illuminated by the sun. As we are told in #D16 'nobody knows what is beyond this'. If we choose the figure of 167 000 *li*, the diameter of this circle is:

$$2 \times 167\,000\ li + 476\,000\ li = 810\,000\ li.$$

This is a highly satisfactory result, for as Zhao Shuang comments:

[144] Compare for instance Liu Xin's attempts to derive the basic constants of his astronomical system from first principles in *Han shu* 21a and 21b.

> [This figure of] 81 is the ultimate Yang number, so it gives the extreme extent of sunlight. (*Zhou bi* 40m in Qian's edn)

The point is that 81 is the square of 9, which is itself the square of the primary Yang number three, corresponding, as we know, to heaven itself. Paragraph #A2 has already stressed the importance of the number 81 as the basis of the order of all geometrical forms. The limit of sunlight is thus numerologically correlated with the furthest extent of Yang, represented by the ultimate number. Of course, had the numbers fallen out differently, the figure of 81 could have been brought in another way. Compare for instance:

> The Yang reaches its culmination at nine. Therefore the circumference of heaven is nine nines which is 810 000 *li.* (*Chun qiu yuan ming bao* 春秋元命包, quoted in *Taiping yulan* 太平御覽 1, 10a–10b)

This is one of the 'apocryphal books' which flourished around the time of Wang Mang and into the Eastern Han.

This does not seem to us the strongest justification for destroying the exactitude of the description of polar conditions already given by Chen Zi. Perhaps he would have retorted that no-one had ever seen the land under the pole, but everybody was agreed about the basic order of the universe as expressed in numbers. Numerology, rational and obvious to all, had to take precedence over what were at most traveller's tales about the far north. At any rate, Chen Zi bequeathed a slightly embarrassing problem to his successors, for now the equinoctial sunlight fell short of the subpolar point by:

178 500 *li* – 167 000 *li* = 11 500 *li.*

Now we recognise this as the radius of the circle described by the *xuan ji* star around the pole. The link with the problem of polar conditions is taken up again in the sequence of paragraphs #F6, #F8, and #F9 (#F7 is clearly a displaced note). It is obvious that for the purpose of the discussion of polar conditions the pole has been replaced by the circumference of the *xuan ji* circle, which *is* just touched by sunlight at the equinoxes as in figure 11 so that:

> #F6 [56f]
> ... the Yang is cut off and the Yin manifests itself.

In his commentary on #B25 Zhao Shuang has already made this point:

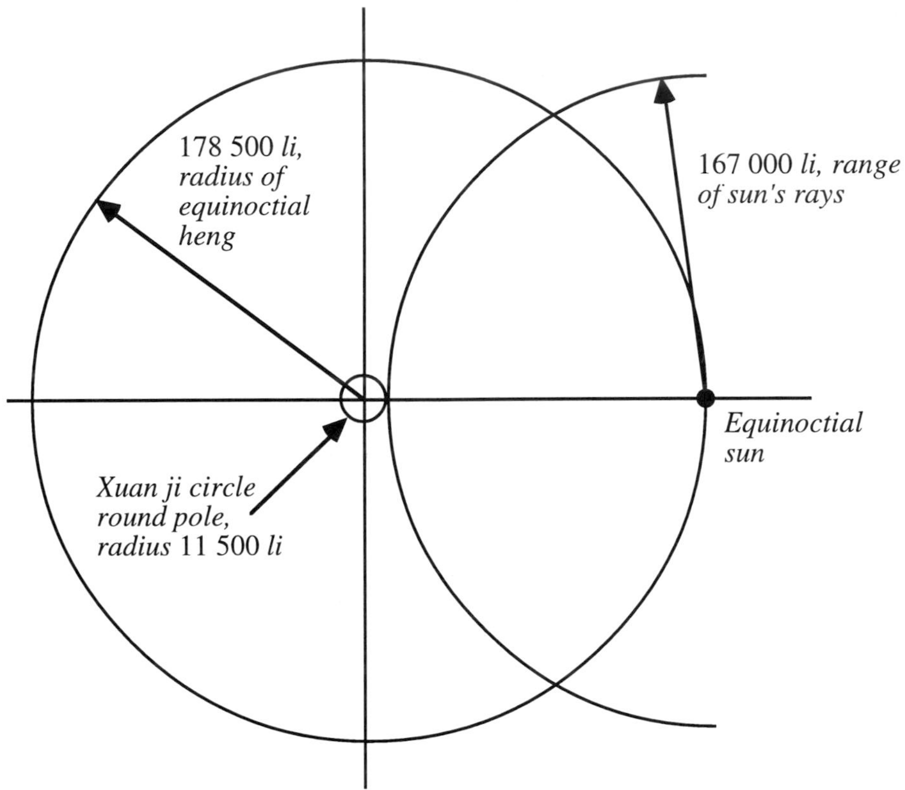

Figure 11. Solar illumination at the equinoxes.

> [To speak of the sunlight] reaching the pole [at the equinoxes] means [reaching] the limit of the *xuan ji* where the Yang is cut off and the Yin manifests itself ... (*Zhou bi* 36p in Qian's edn)

The first problem about the *xuan ji*, its distance from the pole, is now solved. As for the second, its position on what we would now call the solstitial colure between the pole and the winter solstice sun, that question has already been settled in the discussion of the falsified calculations of north polar distances above, where once more the *xuan ji* functioned as a more convenient substitute for the real celestial pole. Having seen that (to use modern terms) both the right ascension and declination of the celestial body described in section #F are quite fictitious, it should be clear that there is absolutely no point in detailed calculations designed to establish the precise epoch at which some real celestial body fulfilled these conditions.[145] All we can say is that the writer of this section of the *Zhou bi*, perhaps genuinely believing that he had discovered important numerical data about the universe, decided that it looked as though the pole star of his day, probably β UMi, might fit the figures he had found, and decided that the theory was so convincing that observational confirmation was superfluous. There is good evidence that the term *xuan ji* was used as a name for the pole star in Western Han times.[146] And thus he created for us a puzzle which has wasted the time of a good number of scholars, both in Asia and the West, but which we can now dismiss as of minor importance.[147] It is, however, another useful link in the evidence enabling us to make some sense of the way the sections of the *Zhou bi* relate to each other.

The sighting tube

There is little to say about this device, which is mentioned only briefly in the dialogue of Chen Zi and Rong Fang:

> #B11 [26h]
> 'Wait until the base is six *chi*, then take a bamboo [tube] of diameter one *cun*, and of length eight *chi*. Catch the light [down the tube] and observe it: the bore exactly covers the sun, and the sun fits into the bore. Thus it can be seen that an amount of

[145] Nōda (1933), 64–98 discusses the *xuan ji* problem in great detail. He finds that in 1100 BC the bright star β UMi had a right ascension of 217.4°, which would have put it almost exactly in the right position between the pole and the winter solstice sun. Unfortunately its declination at that epoch of +83.47° and consequent north polar distance of 6.53° then means that the ends of the cords marking the east–west extremes would have been about 7 *chi* apart rather than the 2.3 *chi* specified in #F2. By the more historically plausible epoch of the beginning of the Western Han this star's right ascension had changed to 242.5°, so that the alignment would no longer work as required. The distance from the pole is also slightly greater, worsening the discrepancy with #F2.

[146] Thus see the gloss of Fu Sheng 伏生 (fl. 200 BC) on a passage in the Book of Documents: 'The thing which changes by a tiny amount but moves something great is called the *xuan ji*; therefore the *xuan ji* [may be] said to be the north pole': *Shang shu da zhuan* 尚書大傳, quoted in *Taiping yulan* 太平御覽 29, 3a. Around 10 BC Liu Xiang said that the *xuan ji* was the 'pivot star' *shu xing* 樞星: *Shuo yuan* 說苑 18, 1b in *Sibu congkan* edn. For more detailed evidence on this point see Cullen (1980), 39–40, which also makes the very tentative suggestion that the *xuan ji* referred to in the Book of Documents might be an early example of the diviner's cosmic model *shi*, with the pole star at the centre of its heaven-disc.

[147] A striking example of the confusion which has been caused is the quite illusory theory of Henri Michel that the *Zhou bi* refers to the use of serrated jade discs as 'circumpolar constellation templates'. This view was unfortunately repeated by Joseph Needham, who reasonably enough was impressed by Michel's apparently conclusive documentation, in fact the result of omission of everything in the text that did not fit his ideas, combined with an inability to read Chinese. See Needham (1959), 336–9. For a discussion of this and related issues, see Cullen and Farrer (1983).

> eighty *cun* gives one *cun* of diameter. ... The oblique distance to the sun from the position of the *bi* is 100 000 *li*. Working things out in proportion, eighty *li* gives one *li* of diameter, thus 100 000 *li* gives 1250 *li* of diameter. So we can state that the diameter of the sun is 1250 *li*.'

The calculations omitted from this quotation were discussed earlier: see page 77ff. Clearly the method makes good sense on the assumptions of the *Zhou bi*.[148] Given a hazy day so that the glare of the sun was reduced, it is quite practical to carry out the sighting procedure described. Once again an alleged empirical result has been shaped by certain guiding assumptions. It is too much of a coincidence that the right length of tube should be exactly the same length as the standard gnomon, and that at the same time its bore should be exactly one *cun*. The figures given here predict an apparent solar diameter of 43', which is about 10' greater than the value actually observed.

The cosmography of the *Zhou bi*

The time has now come to look back over the cosmographical aspects of the *Zhou bi*. I hope that by now it has become evident how difficult it is to separate such questions from the wider astronomical context. The main questions the *Zhou bi* sets out to answer, and the way it answers them, are closely bound up with early Chinese astronomical theory and practice. This was a problem domain in which the main parameter was time rather than space, and this time parameter was mostly measured with respect to events on the observer's meridian line. These included the transits of stars, and the rising and falling of the noon sun throughout the year. The moon was an exception, although as we have seen there are signs that its meridian transit at principal phases in a particular lodge could be used to locate the sun.[149] The *gai tian* 蓋天 universe described earlier fits in well with this way of doing astronomy: as the heavenly disc rotates it displays a sequence of data through the narrow window of the meridian. So long as what is seen through this window is correct, it does not easily occur to the astronomer to ask questions about whether the appearance of the heavens *as a whole* corresponds to the model exactly. Exactitude implies measurement – and measurement is something that only happens on the meridian.

Further, the essential features of the *gai tian* from the point of view of its links with astronomical practice are all contained in its plan view. As in the *shi* cosmic model, which works perfectly well although the heaven-disc is simply placed on top of the earth-plate, questions of the actual height of heaven are not central to the theory. Nor for that matter is the absolute size of the universe. We have seen possible

[148] See Neugebauer (1975) vol. 2, 657–9, for some early Western determinations of the sun's apparent diameter.

[149] *Zhou li zhu shu* 26, 19a, comm.

signs in the third century BC *Lü shi chun qiu* of a version of the *gai tian* in which the use of a different size of gnomon caused all dimensions to be scaled up by a factor of 10/8.

With all this in mind, we will now review the form of the *gai tian* as it appears in the *Zhou bi.* Rather than gathering together every word on the subject from the whole book, it seems better to start by seeing what we can learn from the study of a piece of text that is clearly coherent and self-contained, section #B, which I have called the 'Book of Chen Zi'.

The cosmography of Chen Zi

In the first place it is clear that Chen Zi 陳子 is living in an uncontroversial universe whose general layout needs neither explicit description nor justification. It is obviously what we would now call a *gai tian* universe, but Chen Zi uses no such term. In his intellectual environment conflict between rival views on cosmography has perhaps not yet forced the use of names for different ideas on what heaven and earth are like. Certain minimum features of his cosmography are clear: heaven rotates once daily about the pole, carrying the heavenly bodies with it, and is a uniform height above the earth. We are told nothing explicitly about the shapes of heaven and earth. Taking the earth as flat, which is the normal Chinese view before the seventeenth century, it seems likely that Chen Zi would have held that heaven was flat too.[150] But we must be careful not to force issues that Chen Zi himself does not raise, for by doing so we may give a false picture of his concerns. As we shall see, another section of the *Zhou bi* suggests a different form for heaven and earth.

Chen Zi's real concerns are revealed in the opening question put to him by his disciple Rong Fang

> #B1 [23k]
> Long ago, Rong Fang asked Chen Zi 'Master, I have recently heard something about your Way. Is it really true that your Way is able to comprehend the height and size of the sun, the [area] illuminated by its radiance, the amount of its daily motion, the figures for its greatest and least distances, the extent of human vision, the limits of the four poles, the lodges into which the stars are ordered, and the length and breadth of heaven and earth?'

We have already seen how Chen Zi answers these questions. Apart from the

[150] In defending *gai tian* ideas against what seem to be garbled versions of the *hun tian*, Wang Chong 王充 (AD 27–100) stated that both heaven and earth must be flat and level: *Lun heng* 論衡 11, 10bff., *Hanwei congshu* edn.

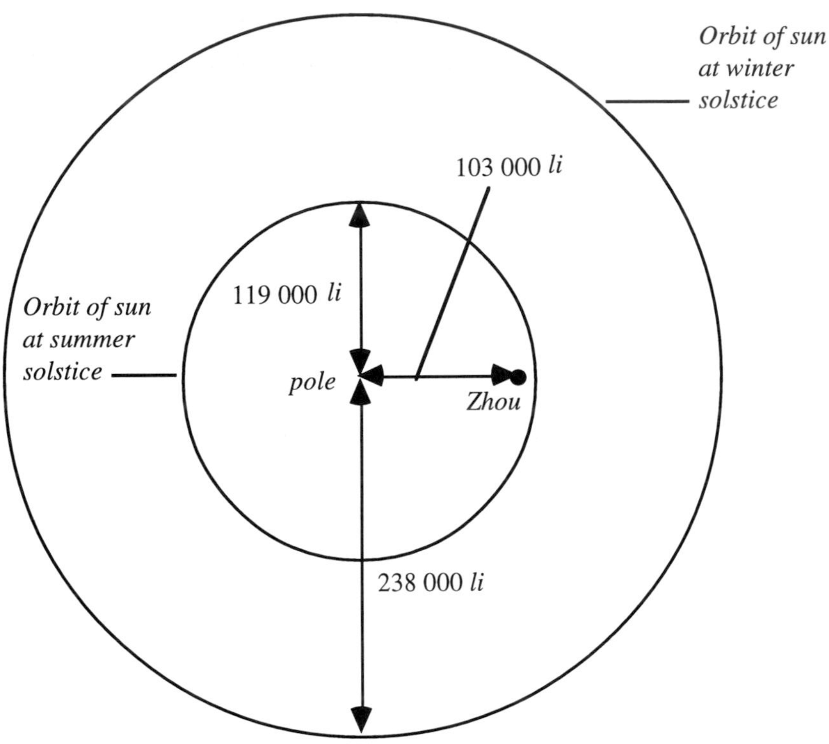

Figure 12. Chen Zi's cosmography, plan view.

height of heaven, all the answers relate to a plan view of the universe, set out here in figure 12. It is a simple and clean view of the cosmos, which does not need to compromise very much with observation to produce such satisfying results as the 1:2 ratio between the summer solstice and winter solstice radius. So far as events on the meridian are concerned, everything functions satisfactorily. As we have seen, the wish to get a numerologically satisfying figure for the ultimate diameter of the universe has meant that a neat account of the six-month alternation of day and night at he pole has had to be complicated somewhat.

Looking at Chen Zi's cosmography overall, some interesting features present themselves. Firstly, the universe is a very large place in comparison with the extent of the Chinese world. The actual distance from the Yellow River down to Canton in the far south (well beyond the pale of civilisation under the Han) is about 2000 *li,* and the circumference of the entire globe is about 47 000 *li.* For Chen Zi, however, the point below the north celestial pole is 103 000 *li* away from his Chinese observer at Zhou, and the diameter of the illuminated region of the earth is 810 000 *li.* This picture of a relatively small China somewhere off–centre on a huge earth is at odds with claims (made elsewhere in classical texts favoured under the Han) that the royal capital was at 'the centre of the land'. The centrality of the royal capital is said in some sources to have been established by gnomon measurements.[151] We know, however, that under the first Qin Emperor the cosmography of Zou Yan was presented to the throne, and this scheme had China well off-centre in the south-east corner of the earth. Such ideas fitted ill with the imperial cosmic synthesis created under the Western Han, and would probably not have been well received. Perhaps the brief change of atmosphere under Wang Mang may have provided a favourable climate for novelties, and as we shall see later there are signs of activity linked with the *Zhou bi* under Wang. Could it be that one of the attractions of the *hun tian* under the Eastern Han was that it restored the Chinese observer to a proper position at the centre of things?

A modern reader is naturally intrigued by those aspects of Chen Zi's world that seem to reflect our present geographical knowledge. I have already mentioned the question of polar conditions several times. There is also the implication that one man's sunset is another man's sunrise, and that our noon is the midnight of an observer on the other side of the pole, points which are developed in later sections, as we shall see. I can think of no likely way in which this view could be the result of actual geographical information. For polar conditions, on the other hand, things are more plausible. There is a continuous land route from the Yellow River to north of the Arctic Circle, and there are other signs that traveller's tales about extreme northern conditions had reached China.[152]

It must be said that even in its own terms the account of Chen Zi has some problems. The use of 167 000 *li* for the extent of the sun's rays rather than 178 500 *li* causes difficulties round the pole, as has already been noted. There is also the fact that although in #B29 Chen Zi calculates the east–west distances through the observer to the solar path at the solstices, he does not do the same for the equinoxes. This would have revealed that when the sun is exactly due east or west at the equinoxes it is 145 172 *li* away, and thus is already within sight range, and therefore has already

[151] See for instance *Zhou li zhu shu* 10, 11b: 'Where the [summer] solstitial shadow of the sun is one *chi* five *cun*, that is called the centre of the earth. It is where heaven and earth join together, where the four seasons succeed one another [rightly], where wind and rain are fitting, and the Yin and Yang are in harmony. Thus everything flourishes in peace. This is where the royal capital is set up. ' More discussion of the idea of Chinese centrality will be found in Cullen (1976), particularly p. 126.

[152] Thus see the story of the 'lamp-dragon' in the northern darkness in *Huai nan zi* 4, 9a, perhaps a reference to the aurora borealis, and the related material quoted in *Chuxue ji* 初學記 3, 60.

risen. Another obvious difficulty is that if heaven rotates at a constant speed, then day and night at the equinoxes cannot be equal if the sun does rise and set due east and west, since less than half of the daily orbit is in sight during the day. Problems of this kind were of course seized upon by later critics of the *gai tian*.[153] Outside the problem domain of meridian observations within which it originated the *gai tian* could give no more than qualitative representations of the phenomena. The *Sui shu* 隋書 put the situation correctly when discussing *gai tu* 蓋圖 'disc-maps [of the heavens]':

> Although they are clear in [predicting the results of] 'leaning back to look upwards' [=meridian observations?], they cannot correctly determine the times of dusk and dawn, or divide the day and night. (*Sui shu* 19, 520)

Other cosmographic material

The subsequent sections of the *Zhou bi* do not change the basic scheme laid down by Chen Zi. They do however add details and attempt to deal with residual problems. Thus section #D does little more than elaborate the data of section #B on the three circles marking the daily paths of the sun at the solstices and equinoxes. In the first place the interval between the solstices is divided more finely, so that a total of seven equally spaced circles now marks the sun's position for twelve of the twenty-four *qi*. Secondly a new term, *heng* 衡, is introduced for these circles, which Chen Zi had simply named *ri dao* 日道 'the path of the sun'. There is no very obvious motivation for the use of this term, which has a range of meanings including 'cross-ways' or 'balance': perhaps there is a reference to the different levels of the noon sun in the sky throughout the year.[154]

It will be recalled that the paragraphs in section #F dealing with the *xuan ji* are motivated partly by the polar problem created by Chen Zi, which they solve by in effect redefining the pole as the region within the *xuan ji* circle of 11 500 *li* radius. It is therefore not unexpected that this section reverts to the topic of polar climatic conditions. We read:

> #F8 [56j]
> The subpolar point does not give birth to the myriad [living] things. How is this known? [When] the winter solstice sun is 119 000 *li* away from the summer solstice [position], the myriad [living] things all die. Now [even] the summer solstice sun is 119 000 *li* away from the north pole. Therefore we know that the subpolar point does not give birth to the myriad [living] things.

[153] See for instance *Sui shu* 19, 506–7.

[154] Or perhaps there is some link with the term *yu heng* 玉衡 'jade *heng*', paired with the expression *xuan ji* 璿璣 in the Book of Documents: see Cullen and Farrer (1983). Since the *Zhou bi* borrows one term, it may have seemed appropriate to find a role for the other.

The point is neatly made. Our winter occurs when the sun moves 119 000 *li* further away than it is at midsummer (when it is only 16 000 *li* away at noon) making a total distance of 135 000 *li*. Now even at midsummer the sun is still 119 000 *li* away from the pole. Clearly nothing like summer can ever occur there at all. But the next paragraph makes an interesting extrapolation:

#F9 [56k]
In the region of the north pole, there is unmelting ice in summer. At the spring equinox, and at the autumn equinox, the sun is on the middle *heng*. From the spring equinox onwards the sun [moves] more and more to the north, and after 59 500 *li* it is at the summer solstice. From the autumn equinox onwards the sun [moves] more and more to the south, and after 59 500 *li* it is at the winter solstice. The middle *heng* is 75 500 *li* from Zhou. Near the middle *heng* there are plants that do not die in winter. [Their situation] is similar to [what it would be if they were] maturing in summer. This means that the Yang manifests itself and the Yin is attenuated, so that the myriad [living] creatures do not die, and the five grains ripen twice in one year. As for the region of the north pole, there are things that grow up in the morning and are gathered in the evening.

Having dealt with the travellers' tales from the far north, the author of this paragraph now rationalises what he has heard about tropical conditions in the far south: we must remember that the tropic of Cancer passes close to Canton. And it must be said that his explanation makes very good sense.

The *xuan ji* material is as we have seen closely related to the work of Chen Zi. Other cosmographical references are more difficult to relate to an overall pattern. They come from section #A, whose language and topic seem to have very little connection with the rest of the book, and from section #E, which seem to be in some disorder. I now discuss the relevant paragraphs, reordering those from section #E so that they read in what seems to be a more rational sequence. The reader who wishes to do so may of course read the text in its original state in the main translation.

#A6 [22p]
'The square pertains to Earth, and the circle pertains to Heaven. Heaven is a circle, and Earth is a square. The numbers of the square are basic, and the circle is produced from the square. One may represent Heaven by a rain-hat. Heaven is blue and black; Earth is yellow and red. As for the numbers of Heaven making a rain-hat, the blue and black make the outside, and the cinnabar and yellow make the inside, so as to represent the positions of Heaven and Earth.'

There is nothing strikingly novel here. We have already seen the way in which the links are made between square/earth/4 and circle/heaven/3. If Song Yu could say that heaven is a cover *gai* over the earth, it seems just as reasonable to call it a rain-hat. The references to colours may be intended to say something about the actual appearance of heaven and earth (blue sky in daytime, black at night, yellow and red soil colours) but they are equally likely to be related to the schemes of cosmic correlation of colours, notes, flavours, directions etc. that reached their peak of integration under the Western Han.[155] It is difficult to see any relation between material of this kind and the cosmography of the rest of the *Zhou bi*. The material from section #E is easier to deal with.

> #E1 [53b]
> The rotation of the sun and moon around the way of the four poles:

> #E4 [53e]
> Therefore when the sun's rotation has brought it to a position north of the pole, it is noon in the northern region and midnight in the southern region. When the sun is east of the pole, it is noon in the eastern region and midnight in the western region. When the sun is south of the pole, it is noon in the southern region and midnight in the northern region. When the sun is west of the pole, it is noon in the western region and midnight in the eastern region.

> #E5 [53g]
> Now in [each of] these four regions heaven and earth have their four poles and their four harmonies. The times of day and night occur in alternation. So in the limits reached by Yin and Yang, and the extremes reached by winter and summer, they are as one.

All this simply makes explicit points that were implicit in Chen Zi's account of the universe, and stresses the cosmic equality of different locations. The last paragraph might even be designed as an explicit rejection of the claims of the *Zhou li* that the Chinese capital was situated at an environmentally perfect position.

> #E6 [54a]
> Heaven resembles a covering rain-hat, while earth is patterned on an inverted pan.

> #E2 [53b]
> As for the subpolar point, it is 60 000 *li* higher than where human beings live, and the pouring waters run down on all sides. Likewise the centre of heaven is 60 000 *li* higher than its edges.

155 On this topic see Graham (1989), 315ff. The Book of Change makes a colour correlation similar to the one found here: 'heaven is black and earth is yellow' *Zhou Yi zhu shu* 1, 27.

#E7 [54b]

Heaven is 80 000 *li* from earth. Even though the winter solstice sun is on the outer *heng*, it is still 20 000 *li* above the land below the pole.

Now this material is both interesting and problematic. The reference to the rain-hat in section #A was clearly unrelated to questions of quantitative cosmography. Here however we have quantitative data, clearly recorded by someone who knows about the *heng* system. But how can we interpret these measurements, and what is their origin? In the first place, there is a problem in interpreting #E6 in conjunction with the other two paragraphs, which seem to imply that heaven and earth both bulge up equal amounts in the middle and are a constant difference apart. A Chinese rain-hat (at least a modern one) is a rather conical affair, whereas all ancient specimens of the vessel type translated here as 'pan' (Chinese *pan* 盤 coincidentally) have a gentle curve from a flattish central base portion. Of course it may be that an ancient Chinese rain-hat may have been more like the type now found in Japan, which is not pointed but gently rounded to make it somewhat pan (or *pan*) shaped.

Turning now to the figures, two problems present themselves. Firstly, it is not easy to make sense of the reference to 'where human beings live'. This point is clearly under the 'edge of heaven' here apparently the outer *heng*, and is not the inhabited area where Chen Zi's observer is placed well within the inner *heng*. Perhaps this phrase should be glossed as '[the most distant place] where human beings live' on the assumption that no-one lives beyond the outer *heng*. Why this limit should be used rather than the outer limit of daylight is unclear.

Secondly, all other distances in the *Zhou bi* have some kind of rational basis within its overall intellectual scheme. It is however very hard to think of any way in which the figure of 60 000 *li* for the bulge could be derived by such means as the 'one *cun* for a thousand *li*' principle. I can only offer a conjecture that there may be a connection with the six intervals, *jian* 間 between the seven *heng*. If each *heng* was 10 000 *li* lower than its predecessor, then the outermost *heng* would indeed be 60 000 *li* lower than the innermost one. I do not pretend to be able to offer any justification for the 10 000 *li* figure; it is however (unlike 60 000 *li*) the sort of figure that could have been arbitrarily chosen. The advantage of confining the fall to the region beyond the first *heng* is that the known world of the Chinese observer is still on comfortingly level ground. The highly conjectural result of this approach is sketched in figure 13.

Interestingly it has a fairly close resemblance to the type of flat-topped *gai* chariot cover often seen in Han bas-reliefs. Other writers have shown heaven and earth as having a gently curving cross-section in which the 'fall' is distributed over the

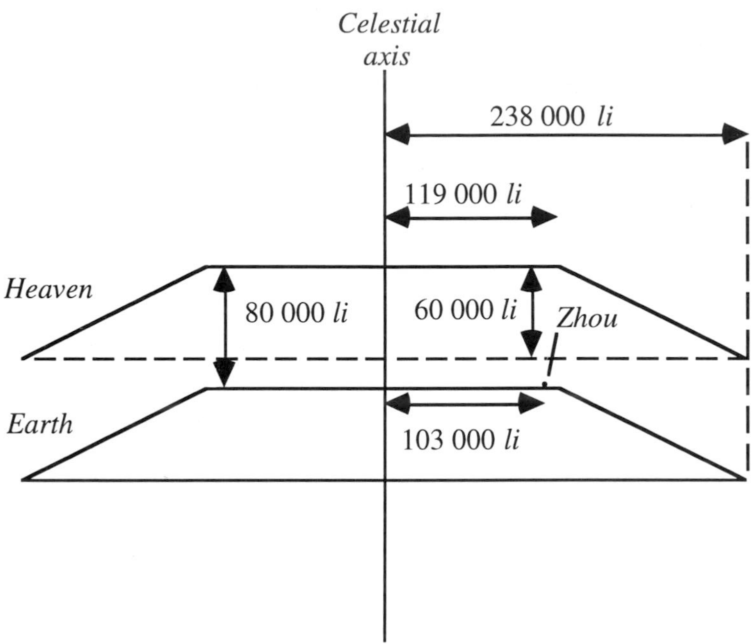

Figure 13. Conjectural restoration of cosmography.

whole of heaven and earth. I feel uncomfortable with the fact that this puts the Chinese observer on a slope. I hope, however, that in producing my tentative sketch of the ideas behind these paragraphs I shall not be said to have produced 'a new interpretation of the *gai tian* theory', for from my point of view such matters are entirely peripheral to the main content of the *gai tian*, which is contained in its plan view.

Now if one is committed to the view that the *Zhou bi* is a single piece of consistent writing, such a reference to heaven and earth bulging upwards is disturbing. In a universe where heaven and earth are not flat and parallel, the shadow rule cannot rationally be applied. In the seventh century Li Chunfeng devoted much effort to pointing this out with detailed arguments.[156] What then becomes of all Chen Zi's reckonings? If however we are not committed to such a view of the text as a unity, the problem dissolves. The author of these paragraphs clearly knew of the system of the

156 See pp. 28–30 in Qian's edition. This material has been discussed in detail in Fu Daiwie (1988).

heng, a term which Chen Zi never uses. He was thus writing at a time when the mathematical core of the *Zhou bi* had already been laid down on the basis of a flat earth and heaven.

If one looks for a reason why such a writer should feel moved to introduce a heaven and earth that were not flat, one can only speculate that he found the idea of a flat heaven somewhat counter intuitive when he looked at the sky apparently coming down towards the earth at the horizon. Perhaps he may have been influenced by contact with *hun tian* ideas? This could have lead him to an arched heaven instead of a flat one, but if so the consciousness that heaven and earth are supposed to be a constant 80 000 *li* apart has evidently led him to give the earth a corresponding bulge. In that case, as with section #G, we have an indication of dating, although in this case comparatively weak. Of course the fact that the lowest part of heaven is still above even the highest regions of earth means that sunrise and sunset etc. will still need to be explained by the use of a fixed range for the sun's rays and human vision, as in #B25.[157]

[157] The *Song shu* is confused when it says that day and night in the *Zhou bi* occur because the sun and moon go out of sight behind the bulge of the earth, quoting a garbled version of the figures in #E2 and #E7 which makes the centre of the earth 20 000 *li* higher than the outer *heng*: see *Song shu* 23, 679.

3

The origins of the work

Most previous discussions of the *Zhou bi* have at some point attacked the apparently obvious question of the date of the work. A recent study cites the opinions of fourteen scholars before coming to its own conclusion that 'Therefore, the date of composition of the work must be in the early Western Han (*c.* 200 BC).' We may note that the opinions cited give dates ranging from the traditional attribution to the Duke of Zhou in the eleventh century BC to the conclusions of a modern scholar that the book can be precisely bracketed into the period AD 9 to AD 84.[158]

I do not intend to provide such a review of the work of all previous scholars on this topic before beginning my own discussion. My reasons for this are twofold. Firstly, and fairly trivially, a critical review of fourteen different opinions (each it must be said written with little or no reference to the others) is probably not the best way to help the reader towards clarity of mind on this question. More importantly, I believe that previous writers on this topic have largely misdirected their efforts in trying to find some epoch when the book as a whole could reasonably have been composed.

I suggest that two unspoken assumptions behind these efforts have lead to largely illusory results being obtained. Firstly, it is assumed that the *Zhou bi* is more or less a single entity which must therefore have some particular date of composition. I believe that on the contrary the different sections of the *Zhou bi* have extremely varied degrees of interrelation. Some seem to have no connection at all, some seem to be later developments of issues raised in earlier sections, and some are flatly inconsistent. What we have here, it seems to me, is a collection of documents made by an individual or by a group with some common interest.[159] If a group was involved, there may have

158 Feng Ligui (1986). I am grateful to Professor Huang Yilong for making this article available to me.

159 Henderson (1991), 35 points to the dangers of exporting the modern concept of a 'book' into the early Chinese situation. He cites work by Fu Sinian, who argues that 'even the concept of the book did not exist in the Zhou [pre-Qin] era, only *p'ien* [*pian*] 篇 (roughly 'chapters'), most of which were on silk rolls or bamboo strips. These *p'ien* were combined and recombined in diverse ways for various reasons to form collections, but not books in the modern sense.' The evidence of some early Han manuscripts, which have chapter titles but no book titles suggest that this practice continued well into the era of interest to us. Keegan (1988) shows clearly how this approach can help us understand the origins of a complex text such as the *Huang di nei jing* 黄帝内經.

been several similar collections differing in material respects. At some point this group, or the member of it whose collection came into the hands of Zhao Shuang, ceased to be active and the collection therefore froze in the state in which it was at that moment. It is illusory to seek for significance in the precise moment when this cessation of activity occurred, or in the precise state of the collection at the moment when it ceased to grow and change. All we can reasonably do is to identify as far as possible the boundaries between the original documents presented as the *Zhou bi*, and then investigate questions of dating separately for each one – so far as that may prove possible.

We now come to the second tacit assumption behind earlier studies. It seems to be taken as axiomatic that the date of composition may be found by setting up a series of dated mathematical and astronomical landmarks against which the *Zhou bi* can be shifted back and forth until a reasonable fit is achieved between its contents and the calibrated background. Firstly the book almost certainly does not have the internal coherence required for this procedure to be meaningful, and secondly we cannot assume that Chinese thought and practice on mathematical and astronomical topics was in any way unified and monolithic during the relevant period.

Arguments of the kind criticised here tend to have the form: 'The *Zhou bi* is more advanced than document X, but it does not yet contain all the features of document Y. Therefore it can be dated somewhere between the two.' Such an argument might work well enough for dating (say) successive maps of a railway system during its period of growth. In the realm of ideas, it will not work except within a single lineage of textual transmission, within which there is likely to be general unity of views at any given period. Without such guaranteed uniformity it is dangerous to assume a unilinear entity called 'Chinese astronomy' against whose development the *Zhou bi* can be calibrated. And where we have clear evidence of strong dissent on astronomical matters, as we do in the Han dynasty, we cannot assume that everybody thought the same way even if they all had equal access to all the relevant information. Under such circumstances the conventional dating procedure loses any value it may have had, particularly in view of the sparseness of datable landmarks in the landscape of the astronomical literature during the period in question.

What follows is therefore not a conventional attempt to 'date the *Zhou bi*'. My analysis will take the following form. Firstly I shall look at the internal structure of the book, and attempt to justify my division of the text into sections. While doing so I shall summarise the few pieces of evidence that may help to relate the material of some sections to datable circumstances. I shall then discuss the interrelations of these sections, and say whatever can safely be said about their individual relations to datable material or events. Next I shall discuss the question of external references (or the lack of them) to the material contained in the collection or to the activities of the putative group. Finally I shall turn to the question of the historical circumstances under which a collection of material like the *Zhou bi* might have been the object of active interest and creativity, and suggest an hypothesis to explain one puzzling feature of the text we have today.

The sections of the *Zhou bi*

Drawing Boundaries

In the main translation, each section is prefaced by an introduction which gives reasons for supposing it to be a coherent block of text, and for separating it from the text blocks which precede and follow it. For clarity I now summarise the evidence on these points. The section titles are my own. The present division into two chapters dates only from the work of Li Chunfeng 李淳風 in the seventh century AD, and has therefore been ignored in the listing.

#A: The book of Shang Gao

There is no difficulty in defining the boundaries of this section, whose style and content differ markedly from the rest of the *Zhou bi*. It opens with the question of the Duke of Zhou, and closes with his expression of satisfaction with the answers he has received from Shang Gao, relating mainly to the trysquare and its mathematical significance, and to numerical lore connected with it.

If we reject the traditional attribution to the beginning of the Western Zhou it is difficult to find internal evidence connecting this section with any particular epoch. Its mathematical content is too elementary for us to be able to set an upper bound on its date on this basis, and its cosmological statements are commonplaces found from the late Warring States onwards. However, as noted earlier (page 89ff.), some of its expressions are closely parallel to those found in an essay by Liu Xin : see *Han shu* 21a, 969–70. It was also pointed out that there was a possible link with references to the circle and square by Liu Xin's father, Liu Xiang. It therefore seems that there would be nothing anomalous in such a text having been composed as late as the early first century AD. The further significance of these textual parallels will be taken up again in chapter 4 (see page 153ff.).

#B: The book of Chen Zi

Like section #A, this section begins with a question, this time from Rong Fang addressed to Chen Zi, in which Rong Fang enquires about the possibility of finding the dimensions of the cosmos by mathematical means. The text which follows has a few relatively insignificant confusions, but continues until all the points asked about by Rong Fang have been met. Like the Duke of Zhou and Shang Gao, Chen Zi and Rong Fang are never mentioned again after their dialogue is concluded. This section could well stand on its own as a systematic exposition of the *gai tian.*

There is little in this section to indicate a date. We have seen signs of a version of the *gai tian* in the *Lü shi chun qiu* in the third century BC, and that text may even show signs of calculating celestial dimensions using the shadow rule but with a gnomon ten *chi* long rather than eight. There are no obvious reasons for setting a lower bound to the composition of this text.

#C: The square and the circle

This fragment hardly deserves to be called a section at all, but stands on its own because it bears little relation to the preceding and following text. On grounds of style and content it might be a displaced fragment of #A. There is nothing to be said about its date, apart from what can be said about #A.

#D: The seven heng

This is a long and well-defined section beginning with the words 'The plan of the seven *heng*'. This is the first time the word *heng* 衡 has appeared in the text: for Chen Zi the sun's daily circles at the solstices and equinoxes were simply *ri dao* 日道 'the solar path'. In addition to creating a new term, this section defines the circles corresponding to a further eight of the twenty-four *qi* 氣, and points out that the sun is half-way between two *heng* for the other twelve. After the main series of calculations ends with #D15 and two paragraphs of comment, #D18 to #D21 add further notes, some of which may be displaced from earlier in the section.

The system of the *heng* is clearly linked to the system of the twenty-four *qi*, and makes no sense without it. We have seen that it is hard to find evidence of the *qi* system earlier than the second century BC. A possible lower bound for the material of this section is provided by a quotation from one of the apocryphal books current in the early years of the first century AD. One of them, the *Xiaojing yuanshen qi* 孝經援神契 'The Book of Filial Piety: a contract which quotes the spirits', was almost certainly forged by a former high official under Wang Mang who tried to make himself emperor in AD 25.[160] One fragment which repeats the figures found in the *Zhou bi* runs:

> As for the circumference of heaven with its seven *heng* and six *jian* intervals, these are 9833 1/3 *li* apart, making 119 000 *li* in all. From the inner *heng* to the middle *heng*, and from the middle *heng* to the outer *heng*, it is 59 500 *li* in each case [reading *ge* 各 for the graphically similar *ming* 名]. (Quoted in *Taiping yulan* 太平御覽 1, 10b)

[160] See Dull (1966), 207–10.

Whoever wrote this was clearly in direct contact with one of the intellectual currents which contributed to the *Zhou bi* as we have it today. It is interesting to see here another sign pointing to connections with Wang Mang's circle.

#E: The shapes of heaven and earth; day and night

The Tang editors divided the originally continuous text at the point where I begin this section. This at least confirms that they saw a natural break in the text where I do, which is helpful given the rather confused nature of this section, in which two sequences of paragraphs seem to be jumbled together. Apart from this, there are other justifications for taking this section as a whole. Firstly, despite its confusion, it does contain fairly coherent statements on the two topics mentioned in my section title. Secondly, it is fairly clear where the preceding and following sections end, and the topics they cover are quite different from those mentioned here. There are no internal indications of dating in this section, apart from the weak suggestion that the bulge of earth and heaven referred to here may be indicative of *hun tian* influence during or after the first century BC.

#F: The xuan ji*; polar and tropical conditions*

This section begins by stating the problem of observing the motions of the *xuan ji* 璿璣 object around the celestial pole, and continues to give practical details of how this is to be done until #F5. The following paragraphs take up the implications of the *xuan ji* radius for polar conditions, and extend the discussion to the tropics. If, as I have argued, we must reject any attempt to date this section astronomically (see page 127) there are no other independent dating indications in the text of this section.

#G: The graduated circle and north polar distances

The beginning of this section is well defined by an abrupt break with the previous topic, and the commencement of instructions for laying out a graduated circle on the ground and using it for observations of the lodges. Further data on the lodges are given in #G11 to #G14, which are also linked with what precedes them by (as I have argued) their tacit aim of competing with *hun tian* procedures. Since the *hun tian* seems to have originated during the first century BC, we have a rough upper bound on the composition of this section. There is also the fact that the lodge Ox is said to have a width of eight *du*. Under the old lodge system apparently in use before 104 BC this lodge had a width of nine *du*: see table 2 (page 18) for the two systems. Once again, section #G seems to come from the first century BC.

#H: The shadow table

Once more the topic changes abruptly. An opening paragraph asks for the shadows at the twenty-four *qi*, the required data are tabulated, and two subsequent paragraphs sketch the underlying calculations. As I have mentioned, the present content of this section is largely the creation of Zhao Shuang himself, who found the older shadow table unsatisfactory: see appendix 4.

Like the system of the *heng*, the structure of the shadow table is bound up with the system of the twenty-four *qi*, which cannot be traced back before the second century BC. As in the old table, the *Huai nan zi* book (*c*. 139 BC) has the *qi* at 15-day intervals rather than a more precise fraction of the 365 1/4 day solar period. It is hard to know how much weight we should place on the anomalous name of the sixth *qi*, which violates a Han taboo on the name of an Emperor who reigned from 156 BC. All these matters have already been set out above (page 108).

#I: Lunar lag

This is a well-ordered and uncorrupt section that obviously stands by itself. Without introduction it launches into a sequence of systematic calculations of changes in the moon's position over various periods of time, and just as abruptly stops. It has no obvious connection with what precedes and follows it. I can think of no useful means of dating this material, which could have been composed at any epoch in which calendrical systems of the quarter-remainder type were in use.

#J: Rising, setting and seasons

After the well-defined ending of section #I we have a rather gnomic series of paragraphs which link the sun's rising and setting positions at the solstices to the twelve 'earthly branches' and the eight trigrams. This material seems to have little direct connection with any other part of the *Zhou bi*, and has no dating indications. Its interest in the use of the trigrams and the branches as horizon divisions may point to a link with the *shi* cosmic model, which is related to the *gai tian* as we have seen.

#K: Calendrical cycles

The section begins with a definition of the month, the day and the year, and then lists five important calendrical cycles. Paragraph #K5 asks a series of questions on the derivation of certain quantities such as the length of the year, and the section (together with the present text of the *Zhou bi*) ends when these questions have been answered. Parallels with the preface of the system promulgated in AD 85 have been noted earlier, but this may mean no more than that both texts are based on quarter-remainder system types. The fact that this section uses the *jian xing* 建星 'Establishment star' as its marker for the winter solstice position is another possible link with the Eastern Han, since Jia Kui claims that this is equivalent to his preferred position in Dipper. But this evidence could equally well put the section earlier, since Jia attributes the Establishment mark-point to four old quarter-remainder type systems current in the early Western Han: see *Hou Han shu*, *zhi* 2, 3027. And to complicate matters still further, the account of the Grand Inception reform in *Han shu* 21a, 975 (but not *Shi ji* 26, 1260) says that its winter solstice marker was the Establishment star, despite Jia Kui. Clearly no reliable dating conclusion can be drawn from all this.

Making connections

At this point the text lies before us broken apart into a number of sections, some with no evident possibility of dating, and some with fairly strong indications of various kinds. Moving on to the next stage, I shall try to see how far it is possible to draw connections between the sections, and examine how far this may be consistent with the dating evidence available.

In the first place, as we have seen, a number of sections seem to have no obvious connection with any of the others, except in the sense that they are all ancient Chinese texts concerned with astronomy and mathematics. In this group I would include #A, #C, the original version of #H before Zhao Shuang's revision, #I, #J and #K.

Of the rest (#B, #D, #E, #F, #G) I am inclined to place #B, 'The Book of Chen Zi' at the head of the rest. If it is removed from the corpus the rest of the related texts are deprived of the essential foundation of the shadow rule, and since some deal with problems unsolved by Chen Zi they lose much of their motivation. Next I think we must place #D 'The Seven *Heng*'. Although Chen Zi evidently has not heard of the *heng*, at least under that name, #D systematises and expands his scheme by bringing it into accordance with the sequence of the twenty-four *qi*. Paragraph #E7 refers to the *heng* and therefore knows of #D, and other parts of this section make statements which draw on #B. With its discussion of the *xuan ji* #F sets out to solve Chen Zi's polar problem, and it also mentions the *heng* system. Section #G is aware of both the *xuan ji* and the *heng*.

It seems therefore that we may have located the core of the *Zhou bi* in this ordered sequence of related sections, which by analogy with certain other ancient Chinese texts we might perhaps call the *nei pian* 內篇 'inner sections'. Section #B is the heart of the work, further developed by #D, and with the three closely related sections #E, #F and #G building upon the first two. Let us briefly recall the sparse dating evidence we have assembled so far in relation to these sections:

#B: Some pre-Qin parallels; no lower bound.
#D: Link with 24 *qi* suggests possibly later than third century BC; a possible quotation *c*. AD 25 sets lower bound.
#E: A weak possibility of *hun tian* influence setting an upper bound of 100 BC.
#F: No dating indications.

#G: Clear influence of *hun tian* and use of new lodge system sets upper bound around 100 BC; no lower bound.

All we can say at this point is that such dating indications as there are do not contradict the logical relations of the sections, or their sequence as they appear in the present text. It would be easy to be tempted to the conclusion that the sections date something like this:

#B: pre-Qin to early Han.
#D: Early Western Han.
#E, #F, #G: First century BC.

But on the evidence so far before us there is no reason at all why the whole collection could not have been written within a single short period at a relatively late date, so long as section #D was available for quotation by AD 25. And of course it is possible that the quotation in the *Xiaojing yuanshen qi* was not directly from the *Zhou bi* at all, but from some other source on which the present text of the *Zhou bi* itself drew later. At this point we certainly cannot claim to have pinned the text down. Let us turn to external evidence to see if it will help us.

Early external evidence

The Han shu *bibliography*

The *Han shu* sets the trend for later standard histories by being the first to include a bibliographical chapter *yi wen zhi* 藝文志. Ban Gu 班固 tells us that he compiled this chapter on the basis of work by Liu Xin 劉欣. Liu's work was a continuation of a project that his father Liu Xiang 劉向 (77–6 BC) had begun in the time of Emperor Cheng 成 (r. 32–7 BC), when an effort was made to collect books from the whole empire. Shortly after Liu Xiang's death came the accession of Emperor Ai 哀 in 5 BC, by which time Wang Mang was already wielding great influence at court. He was an old friend of Liu Xin, and had him promoted to high rank. Under his patronage Liu Xin compiled his 'Seven Summaries' *Qi lue* 七略, based on his father's work.[161]

Ban Gu does not claim to have done any independent bibliographical research, so his monograph represents a summary of the imperial holdings made close to 5 BC.

[161] For Ban Gu's account see the preface to his bibliographical monograph, *Han shu* 30, 1701. Liu Xin's biography also describes his commission from Wang (36, 1967) and his early friendship with the future usurper (36, 1972). After Wang seized power, Liu Xin remained one of Wang's most loyal supporters until late into his reign.

When we turn to this chapter, we look in vain for any title resembling the *Zhou bi*. The nearest we get (which is not very near) is in the section on calendrical astronomy. There we find, in addition to descriptions of various calendrical systems:

> 'The Book of Solar Shadows' *Ri Gui Shu* 日晷: 34 *juan* 卷
> 'The Mathematical Methods of Xu Shang' *Xu Shang suan shu* 許商算術: 26 *juan*
> 'The Mathematical Methods of Du Zhong' *Du Zhong suan shu* 杜忠算術: 16 *juan*
> (*Han shu,* 30, 1766)

It therefore seems probable that the *Zhou bi* was not in the imperial collection when Liu Xin worked on it. Of course this by no means proves that it was not yet in existence in any form.

Yang Xiong, Wang Chong and Cai Yong

We have already seen that sometime around the time when Liu Xin was working on his bibliography Huan Tan 桓譚 was arguing Yang Xiong 楊雄 out of his belief in the *gai tian* (chapter 1). Huan Tan tells us that Yang Xiong

> took it that the heavens were a cover *gai* 蓋 always rotating leftwards [i.e. when one faces the pole], with the sun moon and stars following it from east to west. So he drew a diagram of its form with the graduations of its movements, and checked it against the four seasons, the calendrical constants, dusk and dawn and day and night. (*Taiping yulan* 2, 6b–7a)

There is nothing here to force us to believe that Yang Xiong had seen any of the documents that now make up the *Zhou bi*. On the other hand it does sound as if his diagram had some sort of quantitative basis, in which case he may have produced something resembling the work of Chen Zi or the system of the seven *heng*. The details given are however insufficient for us to say anything definite. This is a pity, for his position at court would have given him easy access to the imperial collection, which was in the charge of his younger contemporary and admirer Liu Xin.

The same cannot be said of Wang Chong 王充 (*c*. AD 27–100), which is again a pity from our point of view, since we can therefore deduce relatively little from his lengthy discussion of issues related to the *gai tian* in his *Lun heng* 論衡 written *c*. AD 83. It is in fact what Wang Chong does not say that is important from our present point of view. He never mentions the term *zhou bi*, and although he refers to a chariot-cover *gai* 蓋 as an analogue for heaven, he never mentions the term *gai tian* 蓋天.[162] He shows no knowledge of Chen Zi's use of gnomons and the shadow rule to

[162] *Lun heng* 11, 8b–9b. It is however interesting that he refers to the illusion that heaven is like an 'overturned bowl' *fu pen* 覆盆 (11, 9b), when in fact it is really flat and level like earth. One is reminded of the reference to earth being like an 'overturned pan' in #E6, although this is probably a coincidence.

establish celestial dimensions.[163] Although he knows the theory that the apparent joining of heaven and earth and the rising and setting of the heavenly bodies are to be explained by the limited range of human sight, he puts no figure to this range, apart from the vague indication that it is to be reckoned in tens of thousands of *li*.[164]

On the whole, all that the evidence of Yang Xiong and Wang Chong can tell us is that the *gai tian* was a live issue in the first century AD. We can deduce nothing about the *Zhou bi* or any part of it. With Cai Yong 蔡邕 the case is different. His memorial of *c*. AD 180 has already been discussed in a different context in chapter 1, but it is worth returning to see it in a different light. It will be recalled that he mentioned the three 'schools' *jia* 家 of those who 'talk about the body of heaven', one of which is called *zhou bi*. However:

> The computational methods of the *Zhou bi* are all extant, but if they are compared with the celestial phenomena they are mostly in error, so the astronomical officials do not use them. (*Hou Han shu*, *zhi* 10, 3217, comm.)

This thoroughgoing rejection gives us our first positive lead. Whereas Yang Xiong and Huan Tan simply refer to 'the *gai tian*' Cai Yong uses a term which links his words firmly with the book of Chen Zi, which is the core of the collection whose history we are trying to trace. The likelihood that he is referring to a document rather than to a set of orally transmitted techniques is made clear by his later lamentations that in a long search of the official archives he had (by contrast, it is implied) failed to find any books giving details of the *hun tian*. And clearly whatever book he saw gave fairly full quantitative details. Given that Zhao Shuang may have been writing his commentary not very long after 180 AD, the best bet seems to be that Cai Yong saw something like the text of the *Zhou bi* we have today.

Dating Zhao Shuang

The nature and contents of Zhao Shuang's 趙爽commentary have already been discussed, principally in chapter 2. My purpose now is to try to set a date for his work, and hence for the closure of the canon of the *Zhou bi*.

At the latest, Zhao Shuang must have written at some time before the middle of the sixth century AD, for that is when the second commentator Zhen Luan 甄鸞

163 Thus he tells us (*Lun heng* 11, 13a–b) that one degree of daily movement by the sun is equal to 2000 *li*. This is not too far from the figure given for a degree on the innermost *heng* in #D8, but it still looks like a guess designed to yield the figure of 1000 *li* in the day and 1000 *li* in the night which he then gives. Wang seems to have no conception of the *heng* system with its degrees of different lengths. The diameter of the sun is given as 1000 *li*, evidently another round figure (11,18a), and incompatible with the degree size given earlier. Heaven is 'more than 60 000 *li* from the earth' (11, 19b).

164 *Lun heng* 11, 10a–b.

worked, and Zhen is clearly writing with Zhao's commentary before him as well as the original text. As we shall see below (page 163), Zhao's commentary is mentioned in the catalogue of the imperial collection of the Sui dynasty (AD 581–618). Zhao's preface refers to Zhang Heng's 張衡 essay *Ling Xian* 靈憲 of around AD 120, so we have a closed time-bracket, although still a rather wide one. Looking through Zhao's commentary, we find in section #K two references to the *Qian xiang* 乾象 system of mathematical astronomy created by Liu Hong 劉洪 near the end of the Han (75f and 79c in Qian's edition).[165] When the Han state fell apart into the Three Kingdoms, this system was in official use in the southern state of Wu 吳 from AD 220 to 280.[166] As a minimum hypothesis we can therefore move the lower bound for Zhao's work to around AD 200, and if we assume he is using the official calendrical system of his time and place, he can be taken as a resident of Wu working sometime in the third century AD. We shall see later (pages 161f.) that Zhao may well have copied from work by Liu Hui that was probably not written until after AD 263. Taking this evidence together with that of Cai Yong, it seems safe to conclude that the *Zhou bi* was not added to after around AD 200.

The formation and closure of the collection

It will be useful to sum up the evidence so far. I have suggested that in the *Zhou bi* we can identify a core of closely related material, the later portions of which cannot be earlier than the first century BC because of the clear influence of the *hun tian*. The collection had not reached the imperial library by around 5 BC, but at some time before his death in AD 18 it seems that Yang Xiong may have had access to similar material. There are strong signs that the book was in existence around AD 200.

What more is there to say? It is worthwhile to consider two further aspects of the origins of the *Zhou bi* which have not usually been discussed. In the first place, how and by whom might such a collection have been formed and used? Secondly, when and under what circumstances might the canon have been closed to produce the work that came into the hands of Zhao Shuang? Let us begin by considering some relevant points about the nature of early Chinese intellectual activity so far as it relates to the production and transmission of texts.

Disciples, databases and early Chinese books

The earliest Chinese references to the terms *gai tian* or *zhou bi* do not associate them with any Chinese equivalent of the word 'theory'. Yang Xiong simply uses *gai tian* on

[165] See *Jin shu* 17, 498.

[166] *San guo zhi* 47, 1129.

its own (*Fa yan* 法言, *Sibu beiyao* edn, 10, 1b). As we have already seen, about AD 180 Cai Yong lists the *zhou bi* as one of the three *jia* 家 (conventionally translated as 'schools') which 'talk about the body of heaven' *lun tian ti* 論天體. What then was a *jia*, and what did it do?

Firstly, as mentioned above, Cai Yong's use of the term *jia* seems to be primarily doxographic. The activity that generated the points of view he mentions is for him essentially something that happened in the past rather than a controversy between living protagonists identified with each *jia*. Had the latter been the case, we might have expected him to name names, but he does not. Of one *jia*, the *xuan ye* 宣夜, he can tell us nothing at all but its title. As for the *zhou bi*, he evidently has seen some documents, for otherwise it is hard to make sense of his claim that its mathematical methods are all extant, *shu shu zhu cun* 數術諸存. The existence of the *jia* of which he approves, the *hun tian*, is only attested by the instruments (and presumably the associated observational and computational procedures?) used by astronomical officials. It is hard to avoid the conclusion that the three entities he groups together are rather different sorts of things.

Is it possible to find any social reality underlying Cai Yong's doxographical analysis? To a large extent it is true that the term *jia* is an 'observer category' rather than an 'actor category'. I cannot myself recall any instance of an early Chinese thinker identifying himself as belonging to a specific *jia*. So far as there is self-identification with a particular group, it is a matter of saying who one's teacher was, and who taught him.[167] It is this master–pupil succession that constitutes the social form of ancient Chinese intellectual activity, and the essential transaction between master and pupil is the transmission of a text, not just in the sense of handing over a book (or more likely allowing one's pupil to make a copy of it) but more importantly in the sense of informing the pupil how the text was to be interpreted and put into practice.

A general presupposition of this transaction is that one's teacher is possessed of authentic knowledge of the texts and their interpretation which others do not have – thus the intellectual lineage defines itself in distinction from those outside it.[168] The text transmitted was not in every case immutable, but might change by growth or

[167] Thus see, for instance, the replies of the early Han physician Chunyu Yi 淳于意 to the interrogatory addressed to him during the political purge after the rebellion of 154 BC: *Shi ji* 105, 2794, 2815 and 2816. Of course it is unlikely that the pattern of training of a Han astronomer was identical to that of a physician.

[168] Continuing the example of Chunyu Yi, we are told that his teacher Yang Qing 陽慶 'passed on to him (*chuan* 傳) the pulse books of Huang Di 黃帝 and Bian Que 扁鵲' *Shi ji* 105, 2794. Later (105, 2810) Chunyu Yi has a confrontation with Sui 遂, a royal physician of the king of Qi 齊, who cites the text of a writing by Bian Que. Chunyu Yi's reply admits that the text does contain those particular words, but stresses that they must be interpreted in the context of proper practice. It is presumably precisely this interpretative method that he would have learned from Yang Qing at the time he was given access to the texts, an advantage which the physician Sui evidently did not have.

attrition in the process of transmission or in response to outside demands.[169] The process was made easier by the fact that most Han 'books' would have consisted of a number of separate rolls of bamboo strips linked by cord,[170] so that the subtraction or addition of a section was not a matter of tearing out pages from a codex, or splicing extra lengths into a scroll. The relevant model is not the Jewish *sofer* counting all the letters in his Torah scroll to make sure he has copied it correctly, but rather a networked computer database to which some have read-only access, but which some privileged users may from time to time be allowed to modify. Perhaps it would be more accurate in the present case to say that they allowed themselves to modify the database, and we must also note that one did not admit openly that the charismatic text solemnly received from one's own teacher had actually *needed* modification. That did not stop modification occurring, however. When such a database ceases to change, it is a sign that the group which used it is no longer active. Similarly, when an ancient Chinese text ceased to change it would be a sign that the lineage whose possession it had been was effectively defunct. At such a time it might be given a written commentary in which later scholars attempted with more or less success to reconstruct the essential interpretative tradition which had once been given orally by masters to their disciples. Later commentators might aim to elucidate or expand on their predecessors. This is what happened to a large number of early texts from the Eastern Han dynasty onwards, and the *Zhou bi* was no exception.[171]

For discussion of a much more complex example of text formation, see the analysis of the *Huang di nei jing* 黃帝內經 'Inner Canon of the Yellow Emperor' corpus in David Keegan (1988). Keegan argues forcibly that the present state of this important text is the result of the grouping and regrouping of a number of shorter documents, some independent of each other, and others designed to explicate or attack earlier texts. Such processes are similar to those which I believe led to the formation of the received text of the *Zhou bi*, and would have taken place at about the same time.

The context of collection

When and where might we find a group which might have written or assembled the

[169] See for instance the account of the development of the Mohist canon in Graham (1978), 22–5.

[170] This physical structure gives rise to a peculiarly Chinese form of bibliographical disaster when the cords holding the strips together perish, and one is left with a disordered heap to be sorted to the best of one's ability and restrung. Such a process leads to the situation where one finds a series of short but fairly coherent passages with little or no logic in their order. It is easy to find such examples in the present text of the *Zhou bi*. Even if this does not happen to the whole roll at once, the end of a roll is peculiarly vulnerable to damage.

[171] For an illuminating account of the origins and functions of commentarial activity in China and the West, see Henderson (1991).

Zhou bi collection? As for the 'when' of this question, since the group postulated must have been active after the rise of the *hun tian* in the first century BC, I am reluctant to go further back than the beginning of the Han for its origins at the very earliest. The *Zhou bi* is not of a length or complexity that requires centuries for its formation. While I agree with Angus Graham that the mathematical traditions of the book may go back to pre-Qin Mohist work of the kind possibly represented in the anomalous addendum to *Huai nan zi* chapter 3, the dialogue of Chen Zi and Rong Fang resembles no Mohist text that has come down to us.[172]

Where in Han society might such a group have operated? The *Zhou bi* is primarily a book about the heavens, and the stereotyped scholarly view is that in China such matters were the province of astronomical officials. Like all stereotypes, this has some truth behind it. The first Western scholars to study the topic were rightly impressed by the fact that Chinese governments had maintained an astronomical establishment over two millennia, in contrast to the situation in the cultures with which they were familiar. The stereotype is reinforced by the fact that the overwhelming majority of our materials on Chinese astronomy are of official origin, in the form of the monographs in the standard histories from the *Shi ji* onwards.

But if the *Zhou bi* was a book produced in official circles, it would surely have been available to Liu Xin when he conducted his bibliographical researches around 5 BC. Further, why is it that the book shows no signs of the constants of the Grand Inception system with its 81 month divisor, which was in official use during the whole of the first century BC? On the contrary, section #K sets out to introduce the beginner to the basic logic of the quarter-remainder system type, a form of calendrical system that was out of official use from 104 BC to AD 85.

In fact it is not difficult to find evidence pointing to considerable astronomical activity outside the specialists of the astronomical establishment throughout much of the Western Han. The first person in charge of calendrical matters under the Han was Zhang Cang 張蒼, a statesman and generalist rather than a technician.[173] In 166–167 BC Zhang's position on calendrical astronomy was successfully attacked by Gongsun Chen, 公孫臣 a commoner from the eastern state of Lu, who was then appointed to high office.[174] In the lead-up to the Grand Inception reform of 104 BC an important role was played by propaganda submitted to the Emperor Wu by Gongsun Qing, 公孫卿 an adventurer in the tradition of *fang shi* 方士 from the old region of Qi 齊 on the north-eastern coastline. He put forward a skilful blend of calendrical mathematics and

[172] See Graham (1978), 371.

[173] *Shi ji* 96, 2681.

[174] *Shi ji* 28, 1381; see also 26, 1260 and 10, 429–30.

suggestions as to how the Emperor might emulate the legendary Yellow Emperor in attaining immortality.[175]

The reform itself involved the assembly of a group which involved officials such as Sima Qian 司馬遷, but also more than twenty calendrical experts 'from among the people' *min jian zhi li zhe* 民間治曆者, including the *fang shi* Tang Du 唐都 who had been the teacher of Sima Qian's father, and Luoxia Hong 落下閎, who we have already heard of as the reputed creator of the *hun tian*.[176] Twenty-seven years after the reform in 78 BC a crisis was caused when the Grand Astrologer (*Tai shi ling* 太史令) himself, Zhang Shouwang 張壽王, submitted a memorial attacking the official system and calling for the adoption (or as he claimed the re-adoption) of the 'Yellow Emperor's calendar'.[177] The matter was only settled after a three-year program of observations and seems to have been connected with the fact noted by Liu Xin that 'Amongst the people there are various [versions of] the Yellow Emperor's calendar' but, he adds with appropriate establishment loyalty 'they are not as perspicuous as those recorded by the astronomical officials.'[178] And finally, matters of calendrical astronomy as well as celestial portents were central to the prolific *chan wei* 讖緯 texts that were widely produced from the time of Wang Mang onwards. One of these, as we have seen, seems to be quoting the *Zhou bi*: see page 141.

Enough has now been said to show that it is easy to find a social basis for the formation of a collection such as the *Zhou bi* in the circumstances of the Western Han, and particularly in the period from about 100 BC onwards when much of the core text of the work must have been written, since only in that period could it have been subject to the influence of the *hun tian*. In the same way that the *Zhou bi* group seems to have taken up a position of opposition to the *hun tian*, it is interesting that it also seems to have resisted calendrical orthodoxy in its partisanship of the quarter-remainder system type. One hesitates a little to make a straightforward link with the partisans of the 'Yellow Emperor's calendar' mentioned by Liu Xin, partly because the *Zhou bi* ascribes the origin of its data to other figures,[179] and partly because its lack of a specified system origin means that the *Zhou bi* does not actually contain all the necessary information for a functioning calendrical system. But the core of the *Zhou bi* was certainly largely the work of a self-consciously partisan group.

175 See Cullen (1993).

176 *Han shu* 21a, 975.

177 *Han shu* 21a, 978.

178 *Hou Han shu, zhi* 2, 3037. Perhaps it is also to Liu Xin that we owe the complaints in the calendrical section of the *Han shu* bibliography, where we read that only the most talented should concern themselves with such matters, and that only trouble will ensue if commoners try to study the heavens: 30, 1767.

179 Section #K speaks of Bao 包 [=Fu 伏] Xi 犧 and Shen Nong 神農 as laying down the basic structure of calendrical astronomy. Both these persons were commonly though to have reigned before the Yellow Emperor.

The context of closure

If we are forced to guess the name of the founder of the group who formed the *Zhou bi*, there are no other obvious candidates apart from Chen Zi 陳子, 'Master Chen' himself. Unlike Shang Gao he does not seem to be intended as a representative of high antiquity chosen to lend the text an air of authority, and his teaching is so prosaic that it seems unreasonable to assume that there was never a Master Chen who taught an equally unfamous Rong Fang. If this is so, it does not seem impossible that Rong Fang himself may have gone on to add section #D with its innovation of the *heng*, and that finally his pupil or pupils built the defensive structure of the remaining core sections *nei pian* as the challenge of the *hun tian* appeared. The last of these, #G is the first in which 'method' paragraphs form an integral part of the text, and it may be that the author of this section added the few such paragraphs found in earlier sections. Finally someone added various pieces of related material to form sections #H to #K. Some of these may have been written by members of the group, and some may have been copied from other sources. On the evidence so far before us, it is hard to tell just when the group ceased to work and the canon was closed. Of course it is quite possible that Master Chen and his disciple are figments of an anonymous author.

There is however a possibility of explaining how the text took on its final form, and it has the merit of providing a role for section #A, which is extremely hard to relate to the likely interests of a group that might have wanted to assemble the other sections of the *Zhou bi*. We have already seen that it makes good sense to see section #B, the 'Book of Chen Zi' as the core of a collection that, naturally enough, consists of a series of documents tagged on after it. But section #A, with which #B has no connection, is at the head of the whole. How can this have come about by any rational process? Books do not, after all, come together without intentional human activity. Someone must have added section #A to the front of the book for some reason. Tentatively, but with the conviction that it will be hard to find a more attractive explanation, I suggest the following process.

The story of Wang Mang's rise to power has already crossed the path of our narrative several times, and we must now return to it. As already mentioned, Wang saw to it that his old friend Liu Xin was appointed to high office around 5 BC. In AD 1, Wang moved a step closer to imperial power when Emperor Ping 平 came to the throne at the age of nine. The young ruler was in fact under the control of the Empress Dowager, of whose clan Wang Mang was a member. The levers of power thus came into his hands, and he became regent with the title of An Han Gong 安漢公 'The Duke bringing tranquillity to the Han'. At his investiture he saw to it that memorials were sent in comparing him to the ancient Duke of Zhou, famed for his selflessness and lack of personal ambition when he was regent in the infancy of King Cheng 成 of

Zhou in the eleventh century BC.[180] At the same time Wang appointed Liu Xin to the new office of Xi-He 羲和, whose title referred to the astronomical officials of the ancient sage Emperor Yao 堯 (see above, page 3). Perhaps in deference to the memory of the Duke, Wang waited until the young Emperor died under mysterious circumstances before forcing another minor heir to abdicate in his favour in AD 9.

In AD 5, the year of the Emperor Ping's death, Wang launched what can only be described as a public relations offensive aimed at recruiting learned men outside official circles to his cause, partly no doubt with the aim of increasing his popularity, and partly because he felt that the new dynasty he was planning would need new expertise. Liu Xin had already supervised the construction of ritual buildings on ancient models:

> So that the Han would be in correspondence with King Wen's [construction of] his Numinous Tower, and the Duke of Zhou's founding of Luo[-Yang]. (*Han shu* 12, 359)

Then, the text continues:

> He summoned those throughout the Empire who had comprehensive understanding of lost classical texts, ancient records, astrology, calendrical reckoning, mathematical harmonics, minor studies, historical documents, arts of all kinds, pharmacognosy, including those who taught the Five Classics, the *Lun yu* 論語 the *Xiao Jing* 孝經, and the *Er Ya* 爾雅. They were to be forwarded through the government relay system in carriages driven [or: provided] by themselves, under a single seal [out of a possible five]. Those who arrived amounted to several thousand persons.

And, we are told elsewhere:

> [Wang Mang] caused the Xi-He official Liu Xin and his colleagues to enter [the conclusions of the meeting] into orderly records and to submit them as memorials; the discussions were highly detailed. Therefore I [Ban Gu] have eliminated heretical expressions, and selected the orthodox meaning, and set them out in the following section. (*Han shu* 21a, 955)

The summary of the conference proceedings occupy most of the rest of the chapter, and end on page 972 with an encomium evidently addressed to Wang Mang, praising him as a sage like Emperor Shun 舜 of old for 'giving a general invitation to the learned, [to conduct] broad deliberations'.

Now as we have already seen, parts of this summary bear a striking resemblance to the cryptic paragraphs of section #A. The reference to 'folding a trysquare' and to the relations between squares and circles are particular examples, but the general tone

[180] *Han shu* 12, 348–9.

of both documents is remarkably similar. In #A both Fu Xi and Yu are mentioned as originators of the mathematical order of the cosmos. We know that Liu Xin gave both of these figures a similarly important role, and believed that at the beginning of the Zhou dynasty the Shang worthy Ji Zi 箕子 had passed on this knowledge to King Wu 武.[181] In a similar way, section #A purports to be the record of a conversation between the Duke of Zhou and another Shang worthy. In general, section #A resembles such parts of the *Han shu* much more than it resembles other parts of the *Zhou bi*. A parallel with expressions used by Liu Xiang about squares and circles was noted earlier.

Given the fact that Wang Mang specifically claimed to be like the Duke of Zhou (not to mention the fact that he clearly wanted to found a new dynasty), things fall into place. Clearly it was an attractive proposition to be summoned to Wang's conference and to be entertained at the public expense, perhaps with the prospect of more lasting benefits to follow if one could make one's talents seem desirable enough. Suppose one had a copy of the 'Book of Chen Zi' and the later material appended to it. Why not see if that would get one invited? Better still, why not preface it with something likely to appeal to what was known to be Wang's preferred self image as the Great Regent, and preferably a little more in tune with the cosmic numerology favoured at court than the rather plain technicalities of the main body of the text? Of course, the fact that the summary of proceedings in the *Han shu* concentrates on matters of mensuration may mean that a potential participant would have in any case have had reason to feel that the relevance of the main conference theme to the 'Book of Chen Zi' was fairly close, since the latter deals with the dimensions of the cosmos. And so the book was brought to Chang'an, presented to the court, and waited in the archives about two hundred years until it was read and contemptuously dismissed by Cai Yong. When it came into Zhao Shuang's hands it had suffered some damage, but not enough to render it unintelligible. Thus it came down to us, headed by its carefully calculated preface which had ensured its preservation in official files.

Some such story is at least a possible explanation of why someone might have added to the front of the main body of the text a preface which refers to the Duke of Zhou, and which undeniably resembles what we know of the doctrines expounded at Wang Mang's AD 5 conference. Presumably potential participants had a fairly clear idea of what was expected of them, and did their best to produce appropriate material. It also explains why Liu Xin had not yet seen the text of the *Zhou bi* when he worked on his bibliography in 5 BC, despite its later currency. I certainly do not claim to have proved that this story is a true one. It is, on the other hand, a plausible explanation of the evidence, and renders the present state of the *Zhou bi* text a lot less mysterious

[181] See the introduction to the monograph on the Five Phases, *Han shu* 27a, 1315. Ban Gu tells us that he is handing on Liu Xin's ideas.

than it would otherwise have been. Under the circumstances it would not be reasonable to expect much more.

4

The later history of the *Zhou bi*

I have now reviewed the context of Han astronomy so far as it relates to the *Zhou bi*, and tried to show how the contents and origin of the book can be understood within that context. From the Han dynasty onwards the nature of our discussion changes, for the *Zhou bi* is no longer the property of an active group of astronomical thinkers, but becomes a classical text subject to the labours of commentators and editors.

So far it has been necessary to be cautious in speaking of 'the *Zhou bi*', for we have had no guarantee that the term referred to anything that had yet taken on a fixed form and content. If my tentative explanation of the process by which the canon was closed is correct, the *Zhou bi* waited at least two centuries before Zhao Shuang saw it, and in that interval it certainly seems that the text suffered some damage and corruption. But a careful reading of the present text and its three commentaries suggests that since the time of Zhao the text has not changed apart from a few minor copyist's errors. The object of this section is to tell the story of how the *Zhou bi* came down to the present day. In the process it will be interesting to say something about its later influence, without becoming too involved in the historical complexities of pre-modern Chinese debates on astronomy and cosmography.

The *Zhou bi* in the period of division

From the break-up of the Han in AD 220 to the reunification under the Sui in AD 581, China underwent three centuries of division and political chaos. These centuries were not 'Dark Ages' like those suffered by Western Europe after the fall of Rome, but the China that emerged from this age was very different from the Han. Amongst other rapid social and cultural changes, the greatest of all was the advent of Buddhism from India, and its successful establishment as a central element in the lives of many Chinese. Regional differences became stronger in the absence of a central authority, and in particular the experiences of China north and south of the Yangzi river diverged markedly. While the north was often dominated by military rulers whose origins lay with the nomadic peoples of the Asian steppes, the south was ruled by a series of aristocratic groups who prided themselves on their role as preservers of ancient Chinese culture, despite their residence beyond the former pale of civilisation marked by the great river. More than ever, we cannot assume that a book was 'known' in China as a whole just because one writer seems to be familiar with it. It is therefore interesting to see what evidence there is of knowledge of the *Zhou bi* at different places and times in this era.

Liu Hui

Our first landmark is of course Zhao Shuang 趙爽, who it seems may have been a man of the southern state of Wu 吳, which existed from AD 222 to 280. We must recall, however, that our only evidence for this is that he twice quotes from the official calendrical system of that state. Now by good fortune we happen to know that the great mathematician Liu Hui 劉徽 wrote his commentary on the *Jiu zhang suan shu* 九章算術 in AD 263, and he appears to have lived in the northern state of Wei 魏.[182] Is there any possibility of getting another 'fix' on Zhao Shuang or the *Zhou bi* through the work of Liu Hui?

Unfortunately no explicit reference to either Zhao or the *Zhou bi* appears in Liu's work, and Zhao refers neither to Liu or to the *Jiu zhang suan shu* itself. This is no surprise, given that these two men may have lived in states that were at war with each other, in an age without any institutions for the systematic dissemination of scientific information. It has however often been argued that there is an obvious connection between their work. Thus for example in their history of Chinese mathematics Li Yan and Du Shiran say:

> because work from [Zhao Shuang's] essay 'Diagrams of base and altitude, circle and square', [see appendix 1] is used in the *Commentary of the Nine Chapters on the Mathematical Art* (for Problems 5 and 11 in the Gougu chapter of the *Nine Chapters* [chapter 9 of the *Jiu zhang suan shu*]) it is possible to say that he lived later than Zhao Shuang. (Li and Du (1987), 65)

If one turns to the text of Liu Hui's commentary, the situation is not quite so straightforward as this might suggest. As I have already said, Liu makes no direct reference to Zhao Shuang's work. Nevertheless, in his explanations of the algorithms which the *Jiu zhang suan shu* gives for the solution of problems involving Pythagoras' theorem Liu refers to diagrams (now lost) which were evidently very similar to those underlying Zhao's essay. An apparently successful attempt to restore the lost diagrams is made by Dai Zhen in his edition of the *Jiu zhang* for the *Siku quanshu*.

182 See *Sui shu* 16, 404. However in his commentary to problem 32 of chapter 1 of the *Jiu zhang suan shu* 九章算術 (p. 193 in the edition of Guo Shuchun (1990)) Liu refers to a bronze measure of the time of Wang Mang stored in the arsenal of the Jin 晉 dynasty, which began in AD 265, so it appears that his work may have been added to under the new government.

Let us look first at the solution to the fifth problem of the Nine Chapters. Since Liu's commentary rather than the main text is our concern, I will save time by saying that this problem amounts to no more than finding the hypotenuse when a base of 20 and an altitude of 21 are given. These numbers have obviously been chosen to make the student's job easier, since they yield a whole-number solution of 29. There is the slight complication that because the 'base' in such problems is conventionally taken as shorter than the 'altitude', the Nine Chapters text takes as 'base' and 'altitude' quantities representing lengths which are in fact respectively vertical and horizontal in the physical situation described, and Liu is concerned to deal with any difficulty this might present to the reader. Part of his commentary runs:

> So we may simply say that the numbers of the two superficies [or 'areas': see below] have simply been switched round within the hypotenuse superficies. We may go on to put one outside the other, with the one situated within forming a square superficies and the one situated outside forming a trysquare superficies. The two outer and inner shapes are transformed but the numbers are balanced. Note further, that in this diagram the trysquare of the base superficies is green, and wraps round the outside of the white [part of the diagram]. This [is the case when] its superficies has the altitude–hypotenuse difference as its breadth, and the altitude–hypotenuse sum as its length, while the altitude superficies is the square within ... [Liu then repeats the description with the roles of base and altitude reversed, and points out that the exchange does not make any difference of principle].(*Jiu zhang suan shu*, 9, 420–1 in Guo Shuchun edn (1990))

Comparison with Zhao Shuang's essay translated in appendix 1 does show several instances of word for word parallels here. One systematic difference is however in the words used for 'area'. Zhao used *shih* 實, whereas Liu uses *mi* 冪, which I render here as 'superficies' to make the distinction clear. Let us turn to the other example mentioned by Li and Du before coming to any conclusions.

The eleventh problem of chapter 9 deals with a door whose height is greater than its breadth by six *chi* 尺 eight *cun* 寸 [= 68 *cun*], and whose diagonal is one *zhang* 丈 [= 100 *cun*]. It is required to find the height and width of the door, which turn out to be 96 *cun* and 28 *cun*, again an instance of a whole number solution arranged for didactic ease. We thus have a problem in which we know the hypotenuse and the difference of the base and altitude, with base and altitude to be found. In the light of the previous parallels, one looks at Zhao's essay for a passage relating to a diagram involving the base–altitude difference, and figure 17 of appendix 1 is the only likely candidate. Interestingly enough, it will be recalled, this is the only diagram to which Zhao actually refers explicitly in his essay. In present texts of the *Zhou bi* this is the 'hypotenuse diagram' *xian tu* 弦圖, the first of the three figures that have apparently been handed down since Zhen Luan. In the light of that diagram Liu Hui's text is easy to follow:

> Referring to the diagram for the positions, the hypotenuse superficies just fills 10 000 [square] *cun*. Double it, subtract the base–altitude difference superficies, divide to open the square. What you get is the number of the sum of height and breadth. Subtract the difference from the sum and halve it [i.e. the remainder], and this is the width of the door. Add the number by which the one is greater than the other [i.e. the given height–breadth difference] and this is the height of the door.
>
> Now this method first finds the half. One *zhang* [i.e. the hypotenuse] multiplied by itself gives four of the red superficies, and one yellow superficies. Halve the difference [of the two sides] and multiply by itself. Now double again {which makes two fourths of the yellow superficies}.[183] Halve the excess, and you have two of the red superficies and one quarter of the yellow superficies. With respect to the large square you have got a quarter. So divide to open the square, and you get half the number of the height–breadth sum. Subtract the half of the difference, and you get the breadth; add, and you get the height of the door.
>
> Further, referring to the areas in the diagram, if you sum the base and altitude [superficies] with one another and add the difference superficies, and then subtract the hypotenuse area, it is [the same as] the accumulation [of the areas previously summed?]. So you first see the hypotenuse, and then know the base and altitude (*Jiu zhang suan shu* 9, 423–4 in Guo Shuchun edn (1990))

Once more, reference to Zhao's essay reveals clear parallels with Liu Hui's text. But who is drawing on whom? We could avoid the difficulty by supposing that both were drawing on some otherwise unknown source, but despite the communication difficulties if Liu and Zhao were approximate contemporaries, mathematical geniuses (like other entities) are perhaps not to be multiplied without good reason. While Li and Du take it that Liu Hui had read Zhao's work, I strongly suspect that if there was influence it was the other way round. My reasons are as follows.

> (1) Zhao's assorted manipulations of areas are almost completely unmotivated by any reference to problems. As such they are anomalous in the history of Chinese mathematics. The only reason he gives us this material seems to be that he wants to add some substance to his unsatisfactory attempts to elucidate the references to trysquares in section #A of the *Zhou bi*. The idea that Zhao summoned these procedures out of thin air as mathematics for mathematics' sake, and that Liu Hui

183 The text is defective here. I follow the emendation of Guo Shuchun given in curly brackets, but it is not easy to see the point of the procedure, nor of the phrase immediately following. Since I am not here engaged in studying Liu's methods for their own sake, I will let the point pass.

later put them to use in relation to the problems in the Nine Chapters strikes me as bizarre.
(2) Liu Hui explicitly refers to the green and white diagrams in his original text (now lost) which were equivalent to the reconstructed figures 19 and 23 which seem to underlie part of Zhao's discussion. Zhao however mentions no diagrams, although they would be obvious aids to someone expounding such matters for the first time. It seems likely that he is simply pillaging Liu Hui's text for his own purposes.
(3) From his voluminous work on the Nine Chapters and from his other mathematical work, it is clear that Liu Hui was a creative mathematician of a high order. Apart from the short essay in question, Zhao shows few other signs of mathematical prowess. I find it more likely that he borrowed from Liu than that a small portion of Liu's work is borrowed from Zhao.
(4) Finally, a significant point emerges from Zhao's essay on *gougu* 勾股 relations. In his commentary on part of the garbled and obscure paragraph #A3 Zhao uses the expression '[the text] is about to apply [this relation] to a myriad things, so here it first sets out the proportions': *jiang yi shi yu wan shi, er ci xian chen qi lu ye* 將以施於萬事而此先陳其率也. In his commentary on the third problem of chapter 9, Liu Hui remarks, referring to the lucid opening sections of the text on which he is commenting '[the text] is about to apply [this relation] to various proportions, so it first writes out these workings to show the source': *jiang yi shi yu zhu lu, gu xian ju ci shu yi jian qi yuan* 將以施於諸率故先具此術以見其源. Clearly borrowing has occurred again. Does it not seem more likely that Zhao, hard pressed to say something relevant about the *Zhou bi*, would borrow from Liu's continuous exposition than that Liu would borrow phrases from Zhao's fragmentary commentary?
(5) Similar remarks apply to Zhao's short essay 'On the height of the sun': see appendix 2. Liu Hui's thoughts are well motivated by his context, whereas Zhao's statements begin and end rather abruptly. Once again, I cannot see why Liu should be supposed to be expanding on Zhao rather than Zhao borrowing a fragment from Liu, and perhaps adding a diagram in accordance with Liu's practice.

If therefore we exclude the possibility of an unknown common source for Liu and Zhao, it seems that the evident copying of one from the other more probably indicates that Zhao had read Liu rather than the reverse.[184] This means that Zhao must have written at some time after AD 263. If we take his references to the *Qian xiang* calendar as indications that he was a man of Wu, we are now in a position to date his

[184] It is notable that Gillon (1977), 289 felt that the heterogeneity of the material in his essay meant that Zhao was acting as an editor rather than an author.

work rather precisely to the seventeen years between this date and the end of Wu in AD 280.

Yu Xi, He Daoyang, Zhao Fei

A few passing references to the *Zhou bi* enable us to trace it through the two centuries after Zhao Shuang. Yu Xi 虞喜 (AD 281–356) and He Daoyang 賀道養 (fl. *c.* AD 420) both comment on the title of the book (*Taiping yulan* 2, 9b and 10a). In AD 437 a copy of the *Zhou bi* in one chapter was presented to Emperor Wen 文 of the Liu Song 劉宋 dynasty in southern China, together with other books from the mathematical astronomer Zhao Fei 趙 䎋(*Song shu* 宋書 98, 2416). Zhao had been in the service of the Hunnish Northern Liang 北涼 dynasty, for whom he had constructed the *Xuan shi* 玄始 astronomical system; his family was from Dunhuang 敦煌 (*Wei shu* 魏書 107a, 2659–60). Zhao is a very common name in Northern China, and in default of other evidence there is no strong reason for believing that Zhao Fei may have been a descendant of Zhao Shuang.

The story of Chenggong Xing

A story from the fifth century shows us that at that time the *Zhou bi* was by no means regarded as an easy book to understand. In the reign of the Emperor Wu 武 (AD 424–452) of the Wei 魏 dynasty which ruled in the north of China there was a famous Daoist adept by the name of Kou Qianzhi 寇謙之. Kou was a friend of the courtier Cui Hao 崔浩, who was skilled in star-lore and mathematical astronomy (*Wei shu* 35, 814–15, 808–9, 825–6). Kou had a hired labourer, Chenggong Xing 成公興, and one day while Kou was sitting under a tree absorbed in calculation Chenggong came to look at what he was doing.

> Kou said to him 'You're only good for muscle-work. Why are you looking at this?' Two or three days later Chenggong came to watch again, and he kept on doing so. Eventually it happened that Kou was doing calculations about the sun, moon and five planets, and there was a problem he could not solve. He lost his temper in frustration. 'What has upset you, sir?' asked Chenggong. Kou replied 'I have studied mathematics for years on end. Recently I have been working through the *Zhou bi* and I can't get things right. That's what has annoyed me. But you don't know anything about this sort of thing, so why do you bother to ask?' Chenggong said 'Try setting out the problem my way sir'. Suddenly the solution appeared; Kou was astonished and did not know what to make of Chengong. He asked to serve him as a pupil. (*Wei shu* 114, 3049–50)

Needless to say, Chenggong turns out to be an immortal in disguise. Whatever the truth behind the story, it makes the point that even a skilled mathematician by the standards of the age was not expected to find the *Zhou bi* plane sailing.

The *Sui shu* catalogue

As we have already seen it was in the next century that Zhen Luan 甄鸞 (fl. AD 560), an official of the Northern Zhou 北周 met an obvious need by providing detailed working for all calculations in the *Zhou bi* itself and in Zhao Shuang's commentary. Unfortunately Zhen does not always seem to have understood every problem himself, and his working has several mistakes.

The *Zhou bi* was, it seems, a fairly well-known book during the period of political division from the third to seventh centuries. The names associated with it are drawn from the states to the north and to the south of the great barrier formed by the Yangzi river. To stand a reasonable chance of survival in a dangerous age, a book needed to achieve a wide distribution. The risks run by books limited to official collections are chillingly illustrated by the fate of the great imperial library formed by the Sui 隋 dynasty (AD 581–618) after it had successfully reunified China. In AD 622 a commissioner was sent by the newly victorious Tang 唐 government to bring the Sui library to the capital. In an accident at the crossing of the Yellow River something like eighty or ninety per cent of the books disappeared into the muddy waters.

Even the catalogue suffered some damage; a revised version has been preserved as chapters 32–35 of the *Sui shu* 隋書, prefaced by a short history of imperial collections before the Tang. This catalogue records three editions of the *Zhou bi:*

> (a) in one chapter, *juan* 卷, with commentary *zhu* 注 by Zhao Ying 趙嬰 [*sic*];
> (b) in one chapter, with detailed working *chong shu* 重術 by Zhen Luan;
> (c) in one chapter, with illustrations *tu* 圖. (*Sui shu* 34, 1018)

Zhen Luan's working presupposes the existence of Zhao Shuang's commentary, and Ying 嬰 is therefore almost certainly a graphic corruption of Shuang 爽. This was the conclusion reached by Bao Huanzhi 鮑澣之 in the postface to his reprinted edition of 1213 (see below). But even if there ever was a commentary by a separate person named Zhao Ying it is now irretrievably lost, and thc qucstion is thus of littlc practical importance. On the question of names, it is clear that the references to 'Zhao Junqing' 趙君卿 in the heading of the work in present editions, and also in the commentaries of Zhen and Li are simply uses of Zhao's 'style' *zi* 字, an additional name assumed on reaching majority. It may well be that the reference to 'illustrations' is an alternative description of the recension with Zhao's commentary, which as we know added diagrams to the text. The literal meaning of the word *juan* conventionally rendered as 'chapter' is 'scroll', and in the age before widespread block printing it is likely that the literal meaning is the accurate one.

Li Chunfeng and the ten mathematical classics

In AD 656 Li Chunfeng 李淳風 (AD 602–670) was instructed to prepare annotated editions of ten mathematical texts for use in the Tang State Academy (*Jiu Tang shu* 舊唐書 19, 2719); amongst these was the *Zhou bi*. Li's annotations take the form of a subcommentary following the commentary of Zhao Shuang and the detailed working of Zhen Luan. In recognition of the *Zhou bi's* new status as a standard text, the words *suan jing* 算經 'mathematical canon' were added to its title. The bibliographical monographs of both standard histories of the Tang continue to list the one-*juan* versions under the names of Zhao Ying and Zhen Luan. However, almost certainly because of its increased bulk, the text with Li's annotations is listed as divided into two *juan* as are all modern texts of the *Zhou bi* (*Jiu Tang shu* 47, 2306; *Xin Tang shu* 新唐書 59, 1543–4 and 1547).

The printed texts

The *Zhou bi* has been relatively fortunate in the circumstances of its transmission: the texts available to us today are still evidently more or less the same as that seen by Zhao. In the case of the *Jiu zhang suan shu* 九章算術 corruption often makes it difficult to follow the details of Liu Hui's commentary, but such problems are rare in the *Zhou bi*. All texts of the *Zhou bi* known today can be traced to three sources:

(1) The first printed edition of the *Zhou bi* (together with eight other mathematical works) issued by the Imperial Library of the Northern Song 宋 dynasty (AD 960–1127) in AD 1084. After the dynasty had been driven out of the north by the Jin 金 Tartars, this edition was reprinted under the Southern Song (AD 1127–1279) by Bao Huanzhi 鮑澣之; his postface bears a date corresponding to 14th December 1213. A copy of Bao's printing survives in the Shanghai library, and has been photographically reprinted by Wenwu chubanshe 文物出版社 (1980) Beijing, with five other works under the overall title *Song ke suan jing liu zhong* 宋刻算經六種. An early Qing tracing of this copy has been photographically reproduced in the *Tianlu linlang congshu* 天錄琳瑯叢書 (1932). The Song printing includes an appendix entitled *Zhou bi yinyi* 周髀音義 'Reading and meaning in the *Zhou bi*' by Li Ji 李籍 an official of the Imperial Library of the Northern Song. It contains notes on the appropriate readings of a number of characters in the text and commentaries, with short discussions of the meanings of the phrases in which

they occur. This appendix is printed with the *Zhou bi* in most *congshu* collections, but is not included by Qian Baocong in his edition: see below.

(2) A text of unknown antecedents printed in AD 1603 by the late Ming bibliophile Hu Zhenheng 胡震亨 in the *Bice huihan* 祕冊彙函 collection. This is the text ultimately followed by a large number of *congshu,* including the *Sibu congkan* 四部叢刊 and *Sibu beiyao* 四部備要.[185] The text has more errors than the Southern Song print. Hu's preface uses the story of Chenggong Xing to argue that special talent is required to appreciate the merits of the book.

(3) A text copied into the *Yongle dadian* 永樂大典 manuscript collection (completed AD 1407), now lost. It was however collated with the *Bice huihan* text by Dai Zhen 戴震 and other editors of the great eighteenth century imperial manuscript collection *Siku quanshu* 四庫全書, and it seems to have been almost identical to the Southern Song printing. At one time the only access to the resultant text was through the versions reprinted in the *Wanyou wenku* 萬有文庫 and *Congshu jicheng* 叢書集成 collections. Now many libraries have copies of the photographic reprint of an entire *Siku quanshu* manuscript by the Commercial Press, Taipei 1983: see vol. 786 for the *Zhou bi*. The *Siku* editors attempt restorations of several of the corrupt or lost diagrams, but without discussion.

A collated edition of the main text of the *Zhou bi* itself is given in Nōda (1933), 141–59 followed by a table of variants. All previous editions have been superseded by the complete critical text of the *Zhou bi* and its commentaries given by Qian Baocong 錢寶琮 in his *Suanjing shishu* 算經十書 (1963), 11–80. Unless otherwise stated, all references to the *Zhou bi* and commentaries in this study are to the page in Qian's edition, with letters to identify columns counted from the right of each page.

Later studies

None of the ancient writers whose work accompanies the text of the *Zhou bi* in standard editions contributes anything indispensable to our understanding of what the book has to say. In the case of some early Chinese books it is at least possible that their commentators were in touch with some version of the tradition of thought that produced the book in the first place. This does not seem to have been the case with Zhao Shuang and his successors. Zhao brings to the text a good understanding of the astronomy and mathematics of his own day, but when the text seems obscure or

[185] For fuller listings of *congshu* in which the *Zhou bi* appears, see Nōda (1933), 5–7, and Qian (1963), 7–9. Now that the Song text, the *Siku quanshu* text and Qian's critical edition are available these editions are of mainly antiquarian interest.

confused to the modern reader Zhao's comments make it clear that he was in no better position than we are today.

While Zhao, Zhen and Li do not solve any problems for us, it would be foolish to ignore the value of their work in showing us how their contemporaries approached a work such as the *Zhou bi,* and thereby providing us with important insights into the development of Chinese astronomy and mathematics. In the case of Zhen Luan it must be admitted that his contribution is simply to demonstrate how very basic were the mathematical attainments of the students for whom he wrote. Zhao and Li are more significant sources. Apart from his running commentary, Zhao's diagrams are of interest as the earliest surviving Chinese examples of mathematical illustration, although by the Song they were already garbled by repeated copying.

Signs of revival of scholarly interest in the *Zhou bi* begin in the late Ming. Hu Zhenheng's reprint of AD 1603 was accompanied by a few short annotations under the name Tang Yin 唐寅, presumably someone other than the well-known poet and painter who lived from AD 1470 to 1524. If Tang ever found time for the austerities of mathematical studies in his otherwise dedicatedly bohemian life, there is no other record of it. It is much less unexpected to find interest in the *Zhou bi* by Zhu Zaiyu 朱載堉 (AD 1536–1611), the Ming prince famous for his work on mathematical harmonics and calendrical astronomy. His short treatise *Yuanfang gougu tujie* 圓方勾股圖解 carries the date 1610; it discusses problems connected with the dialogue of Zhou Gong and Shang Gao in section #A.[186]

Hu Zhenheng's text of the *Zhou bi* was reprinted in the *Jindai bishu* 津逮祕書 collection issued between AD 1630 and about 1642 by Mao Jin 毛晉 (AD 1599–1659), who added a short postface. Mao Jin's youngest son Mao Yi 毛扆 (AD 1640 to after 1710) inherited his father's library and continued his printing business.[187] In AD 1684 he discovered copies of several of the mathematical texts reprinted under the Southern Song by Bao Huanzhi, and had them carefully traced. It was, however, to be two hundred and fifty years before printed copies of these were made (in the *Tianlu linlang congshu*, see above), and indeed it was not until AD 1800 that so much as a list of these rarities was published. Thus when Dai Zhen (AD 1724–1777) collated mathematical texts for the *Siku quanshu* project in the years 1774 to 1776 he does not seem to have been aware of the Song texts of the *Zhou bi* and other works at first hand, although he did have available texts of almost equivalent value from the *Yongle dadian*.

By the time that Dai Zhen worked on his edition of the *Zhou bi* Chinese scholars had already begun to rediscover some of the riches of their country's ancient mathematical

[186] See Goodrich and Fang (1976), 367–71 for a biography of Zhu and references to studies of his work.

[187] Accounts of the Mao's collecting and publishing activities are given in Hummel (1943), 565–6; see also Qian (1963), 7–8.

and astronomical literature. This was a movement begun by Mei Wending 梅文鼎 (AD 1633–1723) and carried further by his grandson Mei Juecheng 梅瑴成 (AD 1681–1763), both of whom were stimulated by contact with the novel scientific methods brought from Europe by Jesuit missionaries from the late Ming onwards.[188] As a result of this stimulus much that had been forgotten was recalled from oblivion. Not only did Chinese scholars come to realise that their own tradition contained much that was equal to or surpassed what had been achieved in the West; in some cases they claimed that what the Jesuits were bringing to China was simply a development based on learning that had travelled westwards from China in previous centuries. Whatever justification there might indeed be for such claims, in the eighteenth century this was more the result of an as yet largely unbroken cultural self confidence than of objective historical study. It is nevertheless interesting to see how things appeared to Dai Zhen in the case of the *Zhou bi.* His judgment of the book runs as follows:

> This work [is based on] the ancient *gai tian* theory. However, from the Han down to the Yuan and Ming dynasties the *hun tian* theory was universally accepted. In the time of the Ming, Europeans entered China and began to make claims that they were expounding new theories. However when they say that the earth is round, this is just [what the *Zhou bi*] means by saying that 'the earth is like an overturned pan' [from which] 'the pouring waters run off on all sides'. When they talk about [the effect of] differences in north–south distances, this is just [what the *Zhou bi*] refers to when it says that 'round the north pole, in summer there is unmelting ice' and that 'near the middle *heng* there are plants that do not die in winter'. This is why the succession of cold and heat differ between north and south. When they talk about [the effect of] differences in east–west distances, this is just [what the *Zhou bi*] refers to when it says that at midday in the east it is midnight in the west, and that at midnight in the east it is midday in the west [so that] 'the times of day and night occur in alternation' like the succession of the four seasons. This is why the times of day at which the inceptions of the twelve nodal *qi* and conjunctions of sun and moon occur differ between east and west. The books of the new mathematical astronomy record that before Tycho [Brahe] the western theory was that the year had 365 1/4 days, so that every four years the remainder amounted to a whole day. This is just [what the *Zhou bi*] refers to as 'three periods of 365 days and one period of 366 days'. Does not the fact that western theories have their origin in the *Zhou bi* exemplify [the saying of Confucius that] 'when the Son of Heaven no longer has [capable] officials, learning will be found to have been preserved amongst, the barbarian nations'? *(Dai Zhen wenji* (1963) 235; this is a slightly shorter version of material in the *Siku quanshu* preface)

[188] On the Meis, see Hummel (1943), 570–1 and 569.

Dai Zhen's interest in the *Zhou bi* predated his work on collating the text by at least twenty years. In 1755 he wrote an essay on the problem of the term *xuan ji* 璿璣 which occurs in section #F, while of his earlier (1752) essay on the geometry of circles and spheres it was said by a contemporary that it 'took the words of the opening section of the *Zhou bi* as a basis, expanded on them, and followed up their full implications.'[189] Two companions of Dai Zhen's early studies, Cheng Yaotian 程瑤田 (1725–1814) and Feng Jing, 馮經 (fl. 1750), also wrote studies of the *Zhou bi.*[190] As in the case of Zhu Zaiyu their interest seems to have been confined to matters connected with the opening dialogue of the book in section #A, then considered the ancient core of the work. They take the opportunity afforded by the exposition of the text to introduce their own mathematical theories at some length. Another contemporary of Dai Zhen was Wu Lang 吳烺 (awarded the *juren* 舉人 degree 1751), eldest son of the famous novelist Wu Jingzi 吳敬梓; he is said to have written a study of the *Zhou bi* which was printed in 1768.[191]

Editions of the *Zhou bi* were published in Japan in 1785, and 1819; the second of these is said to include the earliest known sectional diagram of the universe according to the *gai tian.*[192] This growth of Japanese interest in the *Zhou bi* may perhaps be connected with the Buddhist reaction against Western astronomy around this period.[193]

Gu Guanguang 顧觀光 (1799–1862) wrote textual notes on the *Zhou bi;* these are followed by a postface whose main purpose is to argue that it would be wrong to condemn the *Zhou bi* for putting forward the false view that the earth and the heavens are basically flat rather than spherical. In Gu's opinion the purpose of the *Zhou bi* is to describe a plane polar projection of the heavens rather than their actual physical form. In arriving at this somewhat anachronistic interpretation, he was apparently influenced by his reading of a treatise by Li Zhizao 李之藻 (1569–?), *Hungai tongxian tushuo* 渾蓋通憲圖說, in which Western methods of cartographic projection are outlined.[194]

[189] See *Dai Zhen wenji* (1963), 104–5, 124–9, 223.

[190] For Cheng's relations with Dai, see *Dai Zhen wenji* (1963), 255. Two short works by him, *Zhou bi jushu tuzhu* 周髀矩數圖注 and *Zhou bi yong ju shu* 周髀用矩術 are in the *Gujin suanxue congshu* 古今算學叢書 of 1897. Feng Jing's study *Zhou bi suan jing shu* 周髀算經術 is in the *Congshu jicheng* collection. The preface states that Feng travelled south in Dai Zhen's company, presumably *c.* 1742. A note in Ding and Zhou (1956), 100 says that Feng was awarded the *juren* 舉人 degree in 1770.

[191] *Zhou bi suan jing tuzhu* 周髀算經圖注, see Hummel (1943), 867. Ding and Zhou mention a Japanese reprint dated (1935). I have not seen a copy of this work.

[192] See Nakayama (1969), 288 and 33. I have made no effort to locate these editions, but mention them in case they may turn out to be of some use to a reader with appropriate research interests.

[193] See Nakayama (1969), chapter 15.

[194] See *Zhou bi suan jing jiaokan ji* 周髀算經校勘記 and *Du Zhou bi suan jing shu hou* 讀周髀算經書後, both in *Huailu congshu* 槐廬叢書 (1886). Further textual notes by Sun Yirang 孫詒讓 (1848–1908) are given in his *Zha yi* 札迻 (1894), 11, 5b–8a, facsimile reprint by Shijie Shuju, *Zhongguo xueshu mingzhu* 中國學術名著 series, Taipei 1961.

The eminent mathematician Zou Boqi 鄒伯奇 (1819–1869) wrote a substantial essay on the *Zhou bi* in which he attempted to examine its contents in the context of other evidence relating to ancient astronomy and mathematics. Perceptive as his comments often are, he keeps to the view that the opening section of the text is the basis of the work, and that it was composed shortly after the founding of the Zhou dynasty *c.* 1000 BC. He rejects much of the rest of the book as the work of later scholars who had not understood the theory properly, and suggests corrections calculated on the basis of a spherical earth.

In the present century Qian Baocong 錢寶琮 (1924) is an important discussion of the origin and date of the material assembled in the *Zhou bi.* It is significant that Qian reverses the traditional view put forward by Zou, and claims that the core of the book is section #B, which contains the dialogue of Chen Zi and Rong Fang. Since Qian's discussion of the date of the *Zhou bi* there has been no dissent from his conclusion that much of the book dates from the Western Han dynasty. The critical edition of the text in Qian (1963) is, as already stated, the basis of the present study; it is preceded by a valuable essay on the *Zhou bi,* its commentaries, and bibliographical problems.

The most detailed modern study of the *Zhou bi* is the *Shūhi sankei no kenkyū*周髀算經の研究 (1933) by Nōda Churyō 能田忠亮. Nōda's book stands in the best traditions of Japanese sinology: detailed historical evidence is cited, while the author's scientific training ensures that technical matters are dealt with clearly and concisely. Much of the recent Chinese account of the *Zhou bi* in Chen Zungui 陳遵媯 (1980) 106–187 appears to be drawn directly from Nōda. An excellent annotated translation of the *Zhou bi* into modern Japanese is provided by Hashimoto (1980). The *Zhou bi* is also discussed in the general histories of Chinese mathematics by Qian (1964) 29–31, 57–60, 99–100, and by Li Yan and Du Shiran (1976) 29–39, 70–76, 109. Ding and Zhou (1956) 100 and 528–531 contains a valuable collection of bibliographical material.

As for work in Western languages, a pioneering attack on the *Zhou bi* was made by the young French sinologist Edouard Biot; his (1841) translation was the first to give Westerners access to the text. Although some of Biot's translations and interpretations need revision, his was a very considerable achievement seen against the background of the resources of scholarship available to him. There is a short notice of the *Zhou bi,* with a translation of the opening section, in Wylie (1897) 163–4. Chatley (1938) is a short paper based largely on Biot (1841) and the references to the *Zhou bi* in Maspero (1929). Chatley's discussion of the cosmography of the *Zhou bi* is unfortunately vitiated by an incorrect assumption about the basic form of the *gai tian* universe it describes. Discussions of the *Zhou bi* in the general context of the history of Chinese mathematics and astronomy are to be found in Mikami (1913), Nakayama (1969), 24–35, Martzloff (1988), 113–14, and of course in Needham (1959), principally 19–24 and 210–16.

In this study I am deeply indebted to the scholarship of all who have written on the *Zhou bi* and related topics before me. To avoid a plethora of footnotes, however, I have tried to keep references to secondary sources to a necessary minimum; I hope this will not be interpreted as an attempt to deny other scholars their due credit for priority in putting forward an interpretation or in citing an illuminating text.

Zhou bi suan jing: translation

THE PREFACE OF ZHAO SHUANG 趙爽

Translation

Amongst the high and the great, nothing is greater than heaven, Amongst the deep and the broad, nothing is broader than earth. Their bodies are immense and wide; their shapes extend upwards and outwards through mysterious clarity. One may examine orbital movements back and forth by means of the arcane counterparts [*xuan xiang* 玄象, i.e. the heavenly bodies] but one cannot grasp that vast extent of space directly. One may check on length and shortness by means of the shadow instrument, but the hugeness of things cannot be measured out in graduations.

Even if one exhausts one's spirit in the effort to understand the transformations [of Heaven and Earth], one cannot trace their mysteries to the ultimate limit. One can seek out the obscure and draw out the hidden but still be unable to gain complete knowledge of their subtleties. Thus strange doctrines have been put forward, and dubious principles have been brought forth. Subsequently there came into being the *hun tian* 渾天 and *gai tian* 蓋天. If one takes the two together, one may fill in all the gaps in [one's knowledge of] the way of heaven and earth, and have a means to make visible all the obscurities in them.

The *hun tian* has the text of the *Ling xian* 靈憲 [by Zhang Heng 張衡], while the *gai tian* has the methods of the *Zhou bi* 周髀. Successive ages have preserved them, and the officials have taken them in hand, to enable themselves [like Xi and He in high antiquity] 'in respectful accord with august Heaven, reverently to grant the seasons to the people'. With my mediocre talents and shallow scholarly attainments I [can only regard such matters as if] was looking up at a neighbouring high mountain, and taking respectful note of the track of its shadow. But during a few leisure days of convalescence I happened to look at the *Zhou bi*. Its prescriptions are brief but far-reaching; its words are authoritative and accurate. I feared lest it should be cast aside, [or be thought] hard to penetrate, so that those who discuss the heavens should get nothing from it.

I set out at once, therefore, to construct diagrams in accordance with the text. My sincere hope was to demolish the high walls and reveal the mysteries of the halls and chambers within. Perhaps in time gentlemen with a taste for wide learning may turn their attention to this work.

A. THE BOOK OF SHANG GAO

Introduction

The traditional view represented in the preface of Bao Huanzhi 鮑澣之 dated 1213[195] is that this dialogue records the transmission of a corpus of knowledge from the Shang 商 dynasty to its Zhou 周 successor. According to Bao, the whole book was put together about the time of the Zhou conquest. On a more sceptical view, we may say that whoever placed this section at the head of the present *Zhou bi* text probably intended that his readers should draw some such conclusion. I have suggested in an earlier discussion (page 155) that #A may in fact be a preface added to the book for political reasons near the beginning of the first century AD.

The Duke of Zhou is a well-known historical figure, who played an important role in the consolidation of Zhou power after the conquest of Shang. He was regent during the minority of King Cheng 成 (*c*. 1020–1012 BC), and several chapters of the Book of Documents *Shu Jing* 書經 contain texts of speeches ascribed to him. By Han times he was reputed to have been the author of the *Zhou li* 周禮 ('Ritual of Zhou'), sometimes known as the *Zhou guan* 周官 ('Offices of Zhou').[196] This idealised account of the administrative organisation of the Zhou dynasty contains several mentions of the use of the gnomon shadow for purposes related to the concerns of the *Zhou bi.*[197] It is not therefore surprising to find the Duke appearing to lend his prestige to the present work.

Shang Gao, however, has never been linked with any name from the historical record. According to Zhao Shuang's commentary he was 'a worthy state officer of Zhou times, skilled in mathematics', which adds nothing to what the text says or implies. It seems slightly pointless for the writer of this section to have introduced a fictitious person as the purveyor of mathematical wisdom to the Duke.[198] I suggest that it is possible that Zhao misread his text, so that he read Shang Gao 商高 for an original Shang Rong 商容; the error is graphically likely in the ancient 'seal script'

[195] Not reproduced in Qian's edition, but found in several *congshu* such as the *Sibu beiyao* 四部備要, as well as in the photographic reprint of the Southern Song edition by Wenwu Chubanshe 文物出版社 (1980).

[196] *Shi ji* 33, 1522.

[197] See for instance *Zhou li zhu shu* 周禮注疏 10, 11b.

[198] But see Graham (1989), 215–16 on the practice in ancient China of 'the multiple twist of hiding your pseudonymous work openly under a pseudonym presumed to conceal the identity of someone who while remaining humbly in the shadows taught some famous man the secret of his success'. And in the *Huang di nei jing* 黃帝內經, roughly contemporary with the *Zhou bi*, the Yellow Emperor is given no less than five apparently fictional interlocutors.

form of the characters. If this is so, Shang Rong would be a plausible interlocutor for the Duke. In Western Han times he was said to have been a virtuous minister of the last Shang king, who dismissed him.[199] Other late Shang worthies were Bi Gan 比干, who was murdered, and Ji Zi 箕子, who was imprisoned. After the Zhou victory the new king honoured all three men.[200] Ji Zi is said in the Book of Documents to have given the Zhou king instruction in such matters as divination and the theory of the five phases.[201] While Shang Rong did not enjoy the renown of Ji Zi, he was not an empty name in Han times. A follower of the first Han emperor mentions his case as an example of the recognition of worth,[202] and a work of the second century BC describes his efforts as a peacemaker, and his refusal to accept high office offered by the Zhou conquerors.[203] A work of the third century AD preserves a tradition that he was an expert physiognomist.[204] None of this gives any direct connection with the *Zhou bi*, but it does show that, like Ji Zi, Shang Rong would have seemed a plausible person to have passed on ancient wisdom to one of the founding fathers of the Zhou dynasty.

In his opening question the Duke refers to Bao Xi 包犧 who is more often called Fu Xi 伏犧. This is the first of the mythical Three Sovereigns *San Huang* 三皇 of remote antiquity, to whom is ascribed the origin of human culture. Thus the Great Appendix to the *Yi jing* 易經 'Book of Change' (*c.* 200 BC, see Graham (1989),359) says

> When Bao Xi ruled the world, he looked up and saw signs in heaven, and looked down and saw patterns in earth ... thus he first made the eight trigrams (*Zhou yi zhu shu* 周易注疏, *Xi ci* 繫辭 8, 4b)

In view of the importance of the trysquare in what follows, it is interesting to note that a snake-bodied Fu Xi is shown grasping a trysquare in some Han dynasty tomb reliefs: see for instance the example reproduced in Needham (1954), figure 28.

199 *Shi ji* 3, 107.

200 *Shi ji* 3, *108; 4, 126; 24, 1229.*

201 Book of Documents, chapter 12, *Hong fan* 洪範 'The great plan'.

202 *Shi ji* 55, 2040.

203 *Han shi wai zhuan* 韓詩外傳, quoted in commentary to *Shi ji* 55, 2041.

204 Huangfu Mi's 皇甫謐 *Di wang shi ji* 帝王世紀, quoted in Tang dynasty commentary in *Shu jing zhu shu* 書經注疏, 11, 25a.

Text

#A1 [13b] Long ago, the Duke of Zhou asked Shang Gao 'I have heard, sir, that you excel in numbers. May I ask how Bao Xi laid out the successive degrees of the circumference of heaven in ancient times? Heaven cannot be scaled like a staircase, and earth cannot be measured out with a footrule. Where do the numbers come from?'

#A2 [13f] Shang Gao replied 'The patterns for these numbers come from the circle and the square. The circle comes from the square, the square comes from the trysquare, and the trysquare comes from [the fact that] nine nines are eighty-one.'

#A3 [14b] 'Therefore fold a trysquare so that the base is three in breadth, the altitude is four in extension, and the diameter is five aslant. Having squared its outside, halve it [to obtain] one trysquare. Placing them round together in a ring, one can form three, four and five. The two trysquares have a combined length of twenty-five. This is called the accumulation of trysquares. Thus we see that what made it possible for Yu to set the realm in order was what numbers engender.'

#A4 [22m] The Duke of Zhou exclaimed 'How grandly you have spoken of the numbers! May I ask how the trysquare is used?'

#A5 [22n] Shang Gao said 'The level trysquare is used to set lines true. The supine trysquare is used to sight on heights. The inverted trysquare is used to plumb depths. The recumbent trysquare is used to find distances. The rotated trysquare is used to make circles, and joined trysquares are used to make squares.'

#A6 [22p] 'The square pertains to Earth, and the circle pertains to Heaven. Heaven is a circle, and Earth is a square. The numbers of the square are basic, and the circle is produced from the square. One may represent Heaven by a rain-hat. Heaven is blue and black; Earth is yellow and red. As for the numbers of Heaven making a rain-hat, the blue and black make the outside, and the cinnabar and yellow make the inside, so as to represent the positions of Heaven and Earth.'

#A7 [23f] 'Thus one who knows Earth is wise, but one who knows Heaven is a sage. Wisdom comes from the base [of the right-angled triangle] and the base comes from the trysquare. Through its relations to numbers, what the trysquare does is simply to settle and regulate everything there is.'

#A8 [23j] 'Good!' said the Duke of Zhou.

B. THE BOOK OF CHEN ZI

Introduction

This section records the dialogue of Chen Zi 陳子 'Master Chen' and his disciple Rong Fang 榮方. Neither of these figures is known to history, nor is there any indication of when their conversation is supposed to have taken place, apart from the distancing *xi zhe* 昔者 'formerly' with which it opens.

This is a substantial and coherent body of text, and it is more or less complete in itself. Zhao Shuang, who believed that the *Zhou bi* was a genuine early Zhou work, evidently felt that this section stood out from its surroundings. He says that it is 'not [part of] the original text of the *Zhou bi*', but is the work of scholars writing after the time of the Duke of Zhou.[205] But whereas Zhao wished to exclude this section from what he supposed was the original text handed down by Shang Gao, I suspect that Chen Zi's book may well have been the core 'original text of the *Zhou bi*' round which the rest of the material was assembled. It is, after all, the only section of the extant *Zhou bi* to mention the term which gives the book its title. For argument on this point, see page 144.

The beginning of the text is clearly defined by the introduction of the two new speakers, as well as by changes of style and content. The end is not so unambiguously marked, but it is not a matter of much doubt. In the opening paragraph Rong Fang asks Chen Zi whether it is true that he can provide information on a number of points, and despite some disorder and a few lacunae in our present text it seems clear that Chen Zi is eventually persuaded to continue his explanation until practically all Rong Fang's queries have been answered. Table 5 may help to make this clear.

The final paragraph that is clearly relevant to Chen Zi's exposition is #B34. It consists of what appear to be rough notes of data taken from earlier paragraphs, such as a reader might have jotted down in the space at the end of a book. The beginning of the short section C marks a clear break of topic.

A careful reading of this section suggests that it may be defective or disordered at various points. Thus for instance Rong Fang's query about the meaning of the term *zhou bi* and Chen Zi's answer in #B14 and #B15 seem to come too late. They evidently belong immediately after #B10, in which the term first occurs. If the two relevant paragraphs are moved from their present place in the sequence, #B13 and #B16 link together fairly well. Paragraph #B12 seems to belong with #B10, if indeed it is not simply a reader's note intruded into the text. Other reorderings and excisions might be

205 23k–24a, commentary.

argued for, but since the text is quite comprehensible as it stands a longer discussion would not be justified.

Table 3. The queries of Rong Fang

QUERY	RESPONSE
'the height and size of the sun'	#B11
'the [area] illuminated by its radiance'	#B24 to #B28
'the amount of its daily motion'	Missing
'the figures for its greatest and least distances'	#B16 to #B18
'the extent of human vision'	#B25
'the limits of the four poles'	#B32
'the lodges into which the stars are ordered'	#B20 (some material lost?)
'the length and breadth of heaven and earth'	#B33 (on the assumption that all the earth's surface is sunlit from time to time)

Text

#B1 [23k] Long ago, Rong Fang asked Chen Zi 'Master, I have recently heard something about your Way. Is it really true that your Way is able to comprehend the height and size of the sun, the [area] illuminated by its radiance, the amount of its daily motion, the figures for its greatest and least distances, the extent of human vision, the limits of the four poles, the lodges into which the stars are ordered, and the length and breadth of heaven and earth?'

#B2 [24e] 'It is true' said Chen Zi.

#B3 [24e] Rong Fang asked 'Although I am not intelligent, Master, I would like you to favour me with an explanation. Can someone like me be taught this Way?'

#B4 [24g] Chen Zi replied 'Yes. All these things can be attained to by mathematics. Your ability in mathematics is sufficient to understand such matters if you sincerely give reiterated thought to them.'

#B5 [24j] At this Rong Fang returned home to think, but after several days he had been unable to understand, and going back to see Chen Zi he asked 'I have thought about it without being able to understand. May I venture to enquire further?'

#B6 [24k] Chen Zi replied 'You thought about it, but not to [the point of] maturity. This means you have not been able to grasp the method of surveying distances and rising to the heights, and so in mathematics you are unable to extend categories.[206] This is a case of limited knowledge and insufficient spirit. Now amongst the methods [which are included in] the Way, it is those which are concisely worded but of broad application which are the most illuminating of the categories of understanding. If one asks about one category, and applies [this knowledge] to a myriad affairs, one is said to know the Way. Now what you are studying is mathematical methods, and this requires the use of your understanding. Nevertheless you are in difficulty, which shows that your understanding of the categories is [no more than] elementary. What makes it difficult to understand the methods of the Way is that when one has studied them, one [has to] worry about lack of breadth. Having attained breadth, one [has to] worry about lack of practice. Having attained practice, one [has to] worry about lack of ability to understand. Therefore one studies similar methods in comparison with each other, and one examines similar affairs in comparison with each other. This is what makes the difference between stupid and intelligent scholars, between the worthy and the unworthy. Therefore, it is the ability to distinguish categories in order to unite categories which is the substance of how the worthy one's scholarly patrimony is pure, and of how he applies himself to the practice of understanding. When one studies the same patrimony but cannot enter into the spirit of it, this indicates that the unworthy one lacks wisdom and is unable to apply himself to practice of the patrimony. So if you cannot apply yourself to the practice of mathematics, why should I confuse you with the Way? You must just think the matter out again.'

#B7 [25k] Rong Fang went home again and considered the matter, but after several days he had been unable to understand, and going back to Chen Zi he asked 'I have exerted my powers to the utmost, but my understanding does not go far enough,

[206] That is, think analogically about problem solutions.

and my spirit is not adequate. I cannot reach understanding, and I implore you to explain to me.'

#B8 [25l] Chen Zi said 'Sit down again and I will tell you.' At this Rong Fang returned to his seat and repeated his request. Chen Zi explained to him as follows.

#B9 [26a] '16 000 *li* to the south at the summer solstice, and 135 000 *li* to the south at the winter solstice, if one sets up a post (*gan* 竿) at noon it casts no shadow. This single [fact is the basis of] the numbers of the Way of Heaven.'

#B10 [26c] 'The *zhou bi* 周髀 is eight *chi* in length. On the day of the summer solstice its [noon] shadow is one *chi* and six *cun*. The *bi* 髀 is the altitude [of the right-angled triangle], and the exact [noon] shadow is the base. 1000 *li* due south the base is one *chi* and five *cun*, and 1000 *li* due north the base is one *chi* and seven *cun*. The further south the sun is, the longer the shadow.'

#B11 [26h] 'Wait until the base is six *chi*, then take a bamboo [tube] of diameter one *cun*, and of length eight *chi*. Catch the light [down the tube] and observe it: the bore exactly covers the sun, and the sun fits into the bore. Thus it can be seen that an amount of eighty *cun* gives one *cun* of diameter. So start from the base, and take the *bi* as the altitude. 60 000 *li* from the *bi,* at the subsolar point a *bi* casts no shadow. From this point up to the sun is 80 000 *li*. If we require the oblique distance [from our position] to the sun, take [the distance to] the subsolar point as the base, and take the height of the sun as the altitude. Square both base and altitude, add them and take the square root, which gives the oblique distance to the sun. The oblique distance to the sun from the position of the *bi* is 100 000 *li*. Working things out in proportion, eighty *li* gives one *li* of diameter, thus 100 000 *li* gives 1250 *li* of diameter. So we can state that the diameter of the sun is 1250 *li*.'

#B12 [34h] 'Method: the *zhou bi* is eight *chi* long, and the decrease or increase of the base is one *cun* for a thousand *li*.'

#B13 [34j] 'Therefore it is said: the [celestial] pole is the length and breadth of heaven.'

#B14 [34j] 'Now set up a gnomon (*biao* 表) eight *chi* tall, and sight on the pole: the base is one *zhang* three *cun*. Thus it can be seen that the subpolar point is 103 000 *li* to the north of Zhou 周.'

#B15 [34m] Rong Fang asked 'What is meant by [the term] *zhou bi*?'. Chen Zi replied 'In ancient times the Son of Heaven ruled from Zhou 周. This meant that quantities were observed at Zhou, hence the term *zhou bi*. *Bi* 髀 means *biao* 表, gnomon.'

#B16 [34q] 16 000 *li* south [of Zhou] on the day of the summer solstice, and 135 000 *li* south on the day of the winter solstice, there is no shadow at noon. From this we can see that from the pole south to noon at the summer solstice is 119 000 *li,* and it is the same distance north to midnight. The diameter overall is 238 000 *li,* and this is the diameter of the solar path at the summer solstice. Its circumference is 714 000 *li.*

#B17 [35e] From the summer solstice noon to the winter solstice noon is 119 000 *li,* and it is the same distance north to the subpolar point.

#B18 [35f] Thus from the pole south to the winter solstice noon is 238 000 *li,* and it is the same distance north to midnight. The diameter overall is 476 000 *li,* and this is the diameter of the solar path at the winter solstice. Its circumference is 1 428 000 *li.*

#B19 [35h] From the equinoctial noon north to the subpolar point is 178 500 *li,* and it is the same distance north to midnight. The diameter overall is 357 000 *li,* [and this is the diameter of the solar path at the equinoxes]. Its circumference is 1 071 000 *li.*

#B20 [35m] Therefore it is said: the lunar path always follows the lodges (*xiu* 宿), and the solar path is in exact [correspondence] with the lodges.

#B21 [36f] [If one measures] south to the summer solstice noon and north to the winter solstice midnight, or south to the winter solstice noon and north to the summer solstice midnight, in both cases the diameter is 357 000 *li* and the circumference is 1 071 000 *li.*[207]

#B22 [36h] From the division of day and night at the spring equinox to the division of day and night at the autumn equinox, there is always sunlight at the subpolar point. From the division of day and night at the autumn equinox to the division of day and night at the spring equinox, there is never sunlight at the subpolar point. Therefore, at the time of the division of day and night at the spring and autumn

[207] The anonymous circle whose dimensions are given here corresponds to the great circle of the ecliptic in the *hun tian* scheme. Zhao makes the link explicit in his diagram of the *heng*: see appendix 3.

equinoxes, the area illuminated by the sun extends just up to the pole. This is the equal division of Yin 陰 and Yang 陽.

#B23 [36k] The winter and summer solstices are the greatest expansion and contraction of the [diameter of the] solar path,[208] and the extremes of length and brevity of the day and night. The spring and autumn equinoxes are [the times when] Yin and Yang are equally matched. [As for] the counterparts of day and night, day is Yang and night is Yin. From the spring equinox to the autumn equinox is the counterpart of day. From the autumn equinox to the spring equinox is the counterpart of night.[209]

#B24 [36o] Therefore the illumination of the noon sun at the spring and autumn equinoxes reaches north to the subpolar point, and the illumination of the midnight sun likewise reaches south to the pole. This is the time of the division between day and night.

#B25 [36o] Therefore it is said: the illumination of the sun extends 167 000 *li* to all sides. The distance to which human vision extends must be the same as the extent of solar illumination. The extent of vision from Zhou reaches 64 000 *li* north beyond the pole, and 32 000 *li* south beyond the winter solstice noon point.

#B26 [37h] At noon on the summer solstice the solar illumination extends 48 000 *li* south beyond the winter solstice noon. It extends 16 000 *li* south beyond the limit of human vision, 151 000 *li* north beyond Zhou and 48 000 *li* north beyond the pole.

#B27 [38h] At midnight on the winter solstice the extent of solar illumination southwards falls short of the limit of vision of the human eye by 7000 *li,* and falls 71 000 *li* short of the subpolar point.

#B28 [38n] At the summer solstice the illumination of the sun at noon and the illumination of the sun at midnight overlap by 96 000 *li* across the pole. At the winter solstice the illumination of the sun at noon falls 142 000 *li* short of meeting the illumination of the sun at midnight, and falls 71 000 *li* short of the subpolar point.

#B29 [39d] On the day of the summer solstice, if one sights due east and west of

208 The reference is to the greatest and least diameters of the daily path of the sun round the pole seen against the disc of heaven.

209 On the significance of the term *xiang* 象 'counterpart, model' see Graham (1989), 362–3. For a discussion of the earliest (*c*. 250 BC) long listing of entities in corresponding Yin/Yang pairs see Graham (1989), 330–2.

Zhou then from the subsolar points directly due east and west of Zhou it is 59 598 1/2 *li* to Zhou. On the day of the winter solstice the sun is not visible in the regions due east and west, [however] by calculation we find that from the subsolar points it is 214 557 1/2 *li* to Zhou.

#B30 [40g] All these numbers give the expansion and contraction of the solar path.

#B31 [40h] At the winter and summer solstices, observe the measurements of the pitch-pipes and listen to the sound of the bells. [Do this by] day at the winter solstice [and by] night at the summer solstice.[210]

#B32 [40k] From the extent of the difference of the figures and the limit of solar illumination, the diameter of the four poles is 810 000 *li,* and the circumference is 2 430 000 *li.*

#B33 [40q] From Zhou southwards to the [furthest] place illuminated by the sun is 302 000 *li,* and northwards to the [furthest] place illuminated is 508 000 *li* from Zhou. The distances east and west [from Zhou to the furthest points illuminated] are each 391 683 1/2 *li.* Zhou is 103 000 *li* south of the centre of heaven, and therefore the east–west measurement is shorter than the central diameter by just over 26 632 *li.*

#B34 [42a] 508 000 *li* to the north of Zhou. 135 000 *li* on the day of the summer solstice. The diameter of the solar path at the winter solstice is 476 000 *li,* and its circumference is 1 428 000 *li.* The overall extent of solar illumination measures just over 391 683 *li* east to west through Zhou.

C. THE SQUARE AND THE CIRCLE

Introduction

This short section has little connection with the text which precedes and follows it. If it belongs anywhere else in the *Zhou bi* it would fit quite well with section #A, which is the only other section to mention the trysquare. In current editions #C1 and #C2 are separated by the two diagrams reproduced here as figure 14. According to the text of #C2 the diagram showing the square within the circle should be labelled *yuan fang*

[210] We have a full account of the rituals of listening to the bells and pitch-pipes at the solstices as practised during the Han: see *Hou Han shu* 後漢書, *zhi* 志 5, 3125–6.

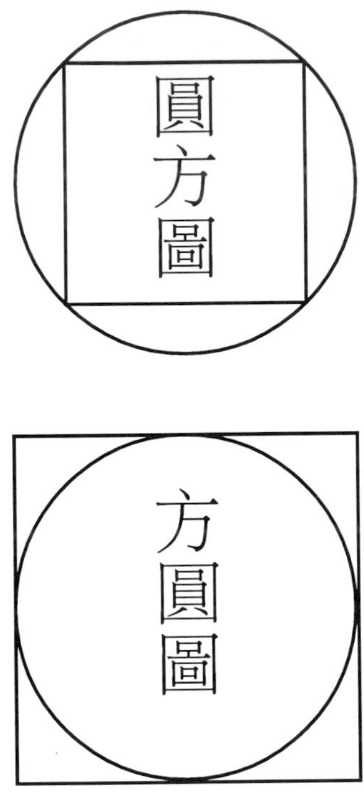

Figure 14. Circle and square.

圓方 'circling the square', while the circle within the square should be labelled *fang yuan* 方圓 'squaring the circle'. In fact the labels are the other way round. This is presumably a slip made by Zhao Shuang when he drew the diagrams, or possibly by a later copyist. The diagrams are given here in corrected form.

Text

#C1 [42d] These are the methods of square and circle.

#C2 [43a] The square and circle are of universal application in all activities of the myriad things. The compasses and the trysquare are deployed in the work of the Great Artificer. A square may be trimmed to make a circle, or a circle may be cut down to make a square. Making a circle within a square is called 'circling the square'; making a square within a circle is called 'squaring the circle'.

D. THE SEVEN *HENG*

Introduction

The main body of this section is a series of calculations repeated more or less identically for eight imaginary celestial circles centred on the pole. The first seven of these circles are the seven equally spaced heng, which represent the daily orbits of the sun round the celestial pole at the dates of the twelve 'medial *qi*' 氣 *zhong qi* 中氣. These are the odd-numbered *qi* in the sequence of 24 equally spaced *qi* which divide up the solar cycle from one winter solstice to the next. The word *heng* 衡 has a range of meanings including 'cross-wise, level'; why it should be used to refer to the solar orbits given here is unclear. Perhaps the reference is to the different levels of the noon sun in the sky at different seasons? The sun's seasonal positions among the *heng* may be shown as in table 4:

Table 4. Relation beween *heng* and *qi*

Heng	*Qi*	
1		13 Summer solstice
	14	12
2	15	11
	16	10
3	17	9
	18	8
4	19 Autumn equinox	7 Spring equinox
	20	6
5	21	5
	22	4
6	23	3
	24	2
7	1 Winter solstice	

At the even numbered *qi*, which are the 'nodal *qi*' *jie qi* 節氣, the sun is half-way between two *heng*. If intercalations are being run properly, the sun should be found on the seventh *heng* at some time during the eleventh lunar month on the Xia 夏 count used today (i.e. with 'Chinese New Year' beginning month #1), on the sixth *heng* during the twelfth month, and so on.

The eighth and outermost circle is not a *heng*, but represents the widest extent of the sun's illumination when it is on the seventh and outermost *heng* at the winter solstice.

The main sequence of calculations begins at #D8, which deals with the innermost *heng*, and continues until #D15 which relates to the limit of illumination. For each of the circles the text gives its diameter, its circumference, and the length of one *du* 度 on this circumference. Paragraphs #D4 to #D7 introduce the *heng* system and explain how the diameter of each *heng* may be obtained. The result for the spacing of the *heng* given in #D7 presupposes knowledge of the total distance between the first and seventh *heng*, a quantity which is not in fact calculated until #D18. I am therefore inclined to move #D18 to between #D6 and #D7, perhaps following it by #D20. Paragraph #D19 seems to be a later note repeating the material of #D15; as in #B32 it appears that the 'four poles' at the edges of the cosmos lie at the limit of solar illumination. The arithmetical explanation of #D20 seems to be a later addition to the text, which otherwise assumes that its readers know how to perform the computations required.

At the beginning of this section the first two paragraphs are clearly linked to a diagram of the seven *heng*. Following the indication in Zhao's preface, the first hypothesis would be that all such diagrams (and hence any explanation of them) must be by Zhao. The situation is however complex. In the first place, as elsewhere, the diagram in all extant editions has clearly been corrupted by an ignorant copyist, since it is quite inconsistent with Zhao's detailed explanation following #D1 (see appendix 3 for translation and discussion). Secondly, in what is now the main text of #D2 we are told of an original large-scale diagram having been halved in size for convenience of representation. The status of this paragraph as part of the main text is apparently confirmed by the fact that it bears a brief comment, presumably by Zhao. It seems, therefore, that we must assume that the text found by Zhao already had a diagram of the *heng* in it. Whether or not he revised this diagram when he copied the text we cannot tell. As we shall see below (section #H) Zhao was prepared to make improvements where he considered them justified.

The quotation from the *Lü shi chun qiu* 呂氏春秋[211] in paragraph #D3 also bears Zhao's commentary, confirming that it was part of the text he saw. He states, however, that it is 'not part of the original *Zhou bi* text'. Although it relates to the general topic of cosmic dimensions, its lack of close connection with its context suggests that it may indeed be an intruded note. Whatever the truth may be, the quotation gives us some rather weak dating evidence: somebody worked on this section, at least to the extent of scribbling a note on it, at some time between the completion of the *Lü shi chun qiu* in 239 BC and whenever it was that Zhao saw the text.

[211] 13, 3a in *Sibu congkan* 四部叢刊 edn.

This section as a whole is evidently not by the same writer as section #B, the 'Book of Chen Zi'. Apart from the fact that the two sections cover much of the same general ground in different ways, Chen Zi never uses the term *heng*. For him the solar orbits are simply *ri dao* 日道 'the path of the sun'. Other sections using the term *heng* are #E, #F, #G and #K, although the last example (#K2) may be no more than intruded text.

Text

#D1 [46a] The plan of the seven *heng.*

#D2 [46j] [Previously] in making this plan, a *zhang* 丈 has been taken as a *chi* 尺, a *chi* has been taken as a *cun* 寸, a *cun* has been taken as a *fen* 分, and a *fen* has been taken as 1000 *li* 里. [So overall] this used [a piece of] silk fabric 8 *chi* 1 *cun* square. Now a piece 4 *chi* 5 *fen* has been used, [so] each *fen* represents 2000 *li.*

#D3 [46l] Mr Lü 吕 said: 'The region within the four seas is 28 000 *li* east–west, and 26 000 *li* north–south'.

#D4 [46q] In their representation of the circles traced by the orbiting sun and moon, the seven *heng* circle round separated by six intervals, which correspond to the six monthly [*qi*] nodes. Six month['s worth of *qi* nodes] make 182 5/8 days.

#D5 [47c] Therefore at the summer solstice the sun is in the lodge Well, and is on the innermost *heng.* At the winter solstice the sun is in the lodge Ox and is on the outermost *heng.*

#D6 [47d] Therefore it is said: a year has 365 1/4 days. In a year [the sun] is once at its nearest distance and once at its furthest distance [from the pole]. In 30 7/16 days the moon is once at its furthest distance and once at its nearest distance [from the pole].[212]

#D7 [47m] Therefore the interval between the *heng* is 19 833 *li* and 1/3 *li,* which is 100 *bu*. To find the diameters of the successive *heng*, double this quantity and add it

[212] The statement about the sun is obvious enough, but the figure given for the moon is odd. 30 7/16 days is simply 1/12 of the solar cycle, and the moon does not of course perform exactly 12 revolutions round heaven (and hence 12 north–south cycles, since its orbit approximates to the ecliptic) for each one of the sun's. The author of section #I gives a clear exposition of such matters. This is another instance suggesting that the *Zhou bi* is a composite work.

to the diameter of the innermost *heng* [to obtain the diameter of the second *heng*]. Add twice as much to the diameter of the innermost *heng* to obtain the diameter of the third *heng*, and so on for successive *heng*.

#D8 [48c] The first and innermost *heng*:
Diameter: 238 000 *li*.
Circumference: 714 000 *li*.
It is divided into 365 1/4 *du*.
One *du* is 1954 *li* 247 *bu* and 933/1461 *bu*.

#D9 [48l] Next is the second *heng*:
Diameter: 277 666 *li* 200 *bu*.
Circumference: 833 000 *li*.
Divide the *li* into *du*:
One *du* is 2280 *li* 188 *bu* and 1332/1461 *bu*.

#D10 [49b] Next is the third *heng*:
Diameter: 317 333 *li* 100 *bu*.
Circumference: 952 000 *li*.
Divide this into *du*:
One *du* is 2606 *li* 130 *bu* and 270/1461 *bu*.

#D11 [49j] Next is the fourth *heng*:
Diameter: 357 000 *li*.
Circumference: 1 071 000 *li*.
Divide this into *du*:
One *du* is 2932 *li* 71 *bu* and 669/1461 *bu*.

#D12 [49p] Next is the fifth *heng*:
Diameter: 396 666 *li* 200 *bu*.
Circumference: 1 190 000 *li*.
Divide this into *du*:
One *du* is 3258 *li* 12 *bu* and 1068/1461 *bu*.

#D13 [50g] Next is the sixth *heng*:
Diameter: 436 333 *li* 100 *bu*.
Circumference: 1 309 000 *li*.
Divide this into *du*:
One *du* is 3583 *li* 254 *bu* and 6/1461 *bu*.

#D14 [50o] Next is the seventh *heng*:
Diameter: 476 000 *li*.
Circumference: 1 428 000 *li*.
Divide this into *du*.
One *du* is 3909 *li* 195 *bu* and 405/1461 *bu*.

#D15 [51c] Next comes the limit of solar illumination at the winter solstice. This goes 167 000 *li* beyond the outermost *heng* [reading *wai* 外 for *bei* 北 unlike Qian and Nōda]. This gives a diameter of 810 000 *li*.
Circumference: 2 430 000 *li*.
This is divided into 365 1/4 *du*.
One *du* is 6652 *li* 293 *bu* and 327/1461 *bu*.

#D16 [51f] Nobody knows what is beyond this.

#D17 [51f] As for those who [claim to?] have some knowledge, some doubt the possibility of knowing anything, while others suspect that the difficulties of knowing [are too great]. This tells us that the highest sage knows without the need for study.

#D18 [51g] Therefore the solar shadow at the winter solstice is 1 *zhang* 3 *chi* and 5 *cun*, and it is 1 *chi* 6 *cun* at the summer solstice. The shadow is longest at the winter solstice and shortest at the summer solstice, changing at the rate of one *cun* for every thousand *li*. Therefore the north–south displacement of the sun between the winter and summer solstices is 119 000 *li*.

#D19 [51j] The diameter of the four poles is 810 000 *li*.
Circumference: 2 430 000 *li*.
Divide this into *du*:
One *du* is 6652 *li* 293 *bu* and 327/1461 *bu*.
This is the distance between *du*.

#D20 [52c] Its north–south displacement in one day is 651 *li* 182 *bu* and 798/1461 *bu*.

#D21 [52e] Method:
Set up the dividend 119 000 *li*.
Take as divisor half a year, 182 5/8 days.
Transform these, obtaining:

Dividend: 952 000 *li*;
Divisor: 1461.
Divide, and the integral quotient is in *li*.
Multiply the remainder by three [and divide], and the integral quotient is in hundreds of *bu*.
Multiply the remainder by ten [and divide], and the integral quotient is in tens of *bu*.
Multiply the remainder by ten [and divide], and the integral quotient is in *bu*.
Make the remainder the numerator of a fraction whose denominator is the divisor.

E. THE SHAPES OF HEAVEN AND EARTH; DAY AND NIGHT

Introduction

It is at this point that the Tang dynasty editors divided the text into two chapters. This break serves to fix the beginning of the present section, and hence also confirms the end-point of section #D. The beginning of section #F is clearly enough defined, but the intervening material seems badly jumbled. Two main topics can be disentangled.

The first of these brings out a conclusion implicit in earlier discussions of the sun's daily orbit round the pole. The universe of the *Zhou bi* has no privileged observers: one observers's noon is midnight for his counterpart on the other side of the pole, and *vice versa*. Paragraphs #E4 and #E5 develop this point, and seem to be in their proper sequence.

The second topic relates to the shapes of heaven and earth, which are discussed briefly in #E2, #E6 and #E7, here separated by other material. The only other material on this topic is the reference in #A6 to heaven resembling a rain-hat. Nothing was said there about earth.

Paragraph #E8 seems unconnected with the rest of the section, and has some affinity with the material in section #J. The reference to the 'four poles' links #E1 with the mention of the outermost limit of daylight in #E3. Both may well belong in the preceding section.

Text

#E1 [53b] The rotation of the sun and moon around the way of the four poles:

#E2 [53b] As for the subpolar point, it is 60 000 *li* higher than where human beings live, and the pouring waters run down on all sides. Likewise the centre of heaven is 60 000 *li* higher than its edges.

#E3 [53d] Therefore the diameter of the outward extent of the sun's rays is 810 000 *li,* and its circumference is 2 430 000 *li.*

#E4 [53e] Therefore when the sun's rotation has brought it to a position north of the pole, it is noon in the northern region and midnight in the southern region. When the sun is east of the pole, it is noon in the eastern region and midnight in the western region. When the sun is south of the pole, it is noon in the southern region and midnight in the northern region. When the sun is west of the pole, it is noon in the western region and midnight in the eastern region.

#E5 [53g] Now in [each of] these four regions heaven and earth have their four poles and their four harmonies. The times of day and night occur in alternation. So in the limits reached by Yin and Yang, and the extremes reached by winter and summer, they are as one.

#E6 [54a] Heaven resembles a covering rain-hat, while earth is patterned on an inverted pan.

#E7 [54b] Heaven is 80 000 *li* from earth. Even though the winter solstice sun is on the outer heng, it is still 20 000 *li* above the land below the pole.

#E8 [54d] Therefore the sun arouses the moon [reading 挑 for 兆], whereupon the moonlight shines forth. The bright moon is formed, and the constellations move in order. Therefore from the autumn equinox up to the winter solstice the essence of the three luminaries weakens, since [the sun's path] is growing more distant. This is the natural working of heaven and earth, and of the Yin and Yang.

F. THE *XUAN JI* ; POLAR AND TROPICAL CONDITIONS

Introduction

This section is principally concerned with an idealised account of gnomon observations of a certain bright circumpolar star, the *xuan ji* 璿璣, whose 'four excursions' are described in #F1. At the end of #F5, it is made clear that the discussion has in effect been completed. Nevertheless, there are good grounds for regarding what follows as part of the same section. The diameter of the circle traced by this star round the pole has been given in #F2; #F6 goes one step further by calculating the circumference, and its last sentence is taken up by #F8, which is clearly linked by topic with #F9. I therefore feel that these three paragraphs cannot easily be separated from the material preceding them, despite the apparent difference of topic. Paragraph #F7 has nothing to do with the present section, and seems to belong in #G: #G2 seems to be a truncated version of this paragraph.

Consideration of the role played by the *xuan ji* in the *Zhou bi* makes it clear why a section on this subject might well contain material of the sort found in the paragraphs on polar and tropical climate. It will be recalled that the definition of this 11 500 *li*–radius circle is a recognition of the anomaly created in section #B, where the 167 000 *li* range given for the sun's rays fell 11 500 *li* short of the range needed for the subpolar point to be illuminated at the equinoxes, as Chen Zi claimed it would be. This section not only makes the discrepancy seem less arbitrary by linking it with the orbit of an important star, but also seems to be attempting to preserve the truth of Chen Zi's claims about six month day and night at the pole. Paragraph #F6 defines the *xuan ji* circle as the borderline between Yang and Yin. Comparing similar phrases in Chen Zi's references to daylight just reaching the pole at the equinoxes, it seems that in effect the *xuan ji* circle is being defined as the border of a circumpolar region. Thus when Chen Zi says that the sun's rays reach the pole at the equinoxes, his statement remains true so long as the sun's rays reach as far as the *xuan ji* circle. In this context it is not surprising that a section commencing with a discussion of the *xuan ji* should go on to discuss the polar climate as in #F8. Tropical conditions, mentioned in #F9, are an obvious comparison.

This is thus apparently the first section of the *Zhou bi* to show signs of critical reconsideration of material occurring earlier in the present text. The tactic of redefining the pole as everything within the *xuan ji* circle seems to lie behind some of the calculations in section #G, which may well be by the same hand.

The *xuan ji* 'star' is in fact an object whose links with observable reality are thin enough to justify applying to it the description 'fictitious'. As already explained, its distance from the pole is simply that required to resolve Chen Zi's polar problem. Its position relative to the sun at the solstices is that required by the false calculation of north polar distances in section #G. Further, even if there had been a celestial body in the correct position, the fact that the solar cycle is not a whole number of days will mean that the neat pattern of positions specified in #F1 will not in fact be seen. For detailed discussion see page 124ff.

Text

#F1 [54j] If you want to know the pivot of the north pole, and the four extremes of the *xuan ji*, you may always take it that:

the time of summer solstice midnight is when the southern excursion of the north pole [star] reaches its maximum;

the time of winter solstice midnight is when the northern excursion reaches its maximum;
the time at the winter solstice when the sun is at *you* 酉 is when the western excursion reaches its maximum;
the time [at the winter solstice] when the sun is at *mao* 卯 is when the eastern excursion reaches its maximum.[213]
These are the four excursions of the *xuan ji* of the north pole.

#F2 [54m] To fix the pivot of the north pole, the centre of the *xuan ji*, to fix the centre of north heaven, to fix the excursions of the [north] pole [star]:

At the winter solstice, at the time when the sun is at *you*, set up an eight-*chi* gnomon, tie a cord to its top and sight [along the cord] on the large star in the middle of the north pole [constellation].[214] Lead the cord down to the ground and note [its position].

Again, as it comes to the light of dawn, at the time when the sun is at *mao*, stretch out another cord and take a sighting with your head against the cord. Take it down to the ground and note [the positions] of the two ends. They are 2 *chi* 3 *cun* apart. Therefore the eastern and western extremes are 23 000 *li* [apart]. The [line of] separation of the two ends fixes east and west, and if one splits [the distance] between them in the middle and points to the gnomon it fixes south and north.

#F3 [55c] The occurrences of these times are all to be checked by waterclock; they are the times of east, west, south and north.

#F4 [55f] [The positions] where it is noted that the cords reach the ground are 1 *zhang* 3 *cun* from the gnomon, and therefore the centre of heaven is 103 000 *li* from Zhou.

#F5 [56a] How do we know the times of the southern and northern extremes? From the fact that the northernmost excursion at midnight on the winter solstice goes 11 500 *li* beyond the centre of heaven, and that the southernmost excursion at midnight on the summer solstice is 11 500 *li* nearer us than the centre of heaven. All this is found by taking sights with the cord tied to the top of the gnomon.

[213] In the *gai tian* 蓋天 universe as represented by the *shi* 式 diviner's cosmic model discussed in chapter 1 the sun is at the positions marked by the cyclical signs *you* and *mao* (due east and west of the pole) at the central instants of the double-hours which are also so named (5–7 a.m. and 5–7 p.m.).

[214] The constellation known since Han times as *Bei ji* 北極 'north pole' is reckoned to consist of a line of stars running from γ and β Ursae Minoris to Σ 1694 Camelopardalis. The brightest star included is β UMi.

For the northern extreme this is observed to reach the ground 1 *zhang* 1 *chi* and 4 1/2 *cun* [from the gnomon], so this is 114 500 *li* from Zhou, and goes 11 500 *li* beyond the centre of heaven. For the southern extreme it is observed to reach the ground at 9 *chi* 1 1/2 *cun*, so this is 91 500 *li* from Zhou and is 11 500 *li* nearer us than the centre of heaven.

This is the method for the four extremes of the *xuan ji*, and for its south and north positions beyond and short of the pole, and the true bases [of right-angled triangles] for east, west, south and north.

#F6 [56f] The diameter of the *xuan ji* [circle] is 23 000 *li* and its circumference is 69 000 *li*. This [means that] the Yang is cut off and the Yin manifests itself [within this region], so that it does not give birth to the myriad [living] things.

#F7 [56g] The method says:
Fix it by setting up the true [shadow] base.
When the sun first rises, set up a gnomon and note its shadow.
When the sun sets, note the shadow again.
The line between the two ends fixes east and west, and if one splits [the distance] between them in the middle and points to the gnomon it fixes south and north.

#F8 [56j] The subpolar point does not give birth to the myriad [living] things. How is this known? [When] the winter solstice sun is 119 000 *li* away from the summer solstice [position], the myriad [living] things all die. Now [even] the summer solstice sun is 119 000 *li* away from the north pole. Therefore we know that the subpolar point does not give birth to the myriad [living] things.

#F9 [56k] In the region of the north pole, there is unmelting ice in summer. At the spring equinox, and at the autumn equinox, the sun is on the middle *heng*. From the spring equinox onwards the sun [moves] more and more to the north, and after 59 500 *li* it is at the summer solstice. From the autumn equinox onwards the sun [moves] more and more to the south, and after 59 500 *li* it is at the winter solstice. The middle *heng* is 75 500 *li* from Zhou. Near the middle *heng* there are plants that do not die in winter. [Their situation] is similar to [what it would be if they were] maturing in summer. This means that the Yang manifests itself and the Yin is attenuated, so that the myriad [living] creatures do not die, and the five grains ripen twice in one year. As for the region of the north pole, there are things that grow up in the morning and are gathered in the evening.[215]

[215] I read *huo* 穫 'harvest, gather' with the 'grain' radical as suggested by Zhao rather than 獲 with the 'dog' radical as in the text, meaning 'get, obtain'. The reference is to the 6-month day/night alternation

G. THE GRADUATED CIRCLE AND NORTH POLAR DISTANCES

Introduction

The introduction of the *xuan ji* circle was part of an effort to face up to a problem posed within the vocabulary of the *gai tian* 蓋天 cosmography. In this section an attempt is made to show that questions which can, in reality, only be asked and answered within the doctrinal scheme of the *hun tian* 渾天 can be solved by *Zhou bi* methods. In #G1 to #G5 the writer describes an observational procedure which can allegedly determine differences of what we would call right ascension without the use of an armillary ring or a timing device. Paragraphs #G6 to #G8 extend the discussion to a problem related to the *shi* 式 cosmic model. The rest of the section gives calculations by which the north polar distances of the sun at the solstices and equinoxes are supposedly derived from quantities based on the mathematics of the *Zhou bi*, again without the need for actual observation using armillary devices. For discussion see chapter 2.

The whole section seems to be in a good state of preservation, apart from #G2, of which a fuller version is apparently preserved in #F7. Apart from an evident unity of style, the material given here is clearly tied together by its polemical purpose of upholding the credit of the methods of the *Zhou bi* against the assaults of the *hun tian*. The fact that the attempt to derive north polar distances can only succeed by more or less *ad hoc* use of the *xuan ji* radius links this section closely with #F.

Text

#G1 [58b] Set up the degrees of the 28 lodges, using the pattern of the successive degrees of the circumference of heaven.

#G2 [58c] Method:
Turn the back on (?) the exact south; fix it by checking the [shadow] base.

#G3 [58d] First level some ground 21 *bu* in diameter and 63 *bu* in circumference, and make it exactly flat using a water level. Then set up a diameter of 121 *chi* 7 *cun* 5 *fen*; multiplying this by three makes [a circumference of] 365 1/4 *chi*, which corresponds to the circumference of heaven, 365 1/4 *du*. Divide it up carefully, so that there is not

near the pole, where spring is dawn and autumn is dusk. I omit *dong sheng zhi lei* 冬生之類 which Qian supplies from Zhao's commentary. This phrase appears in no edition of the main text, and does not seem essential to the sense.

the slightest discrepancy. When this has been done, then fix the north–south and east–west diameters, dividing it into four so that each part is 91 *du* and 5/16 *du*.

#G4 [58n] With this the circle is fixed and true. Next set up a gnomon in the exact centre of north and south, tie a string to its top, and sight on the centring of the middle star of [the asterism] Ox. Next further observe the leading star of [the asterism] Woman.[216] As previously, sight with the gnomon and string on the leading star of Woman, so as to fix its [moment of] centring. Thereupon, using a displacement marker, sight [again] on the middle star of Ox, [noting] how many *du* it is to the west of the central standard gnomon. In every such case, the number of *chi* indicated by the marker gives the number of *du*. The marker will be over the eight *chi* [mark], so we know that Ox has eight *du*. Proceed accordingly with succeeding stars, so that all the 28 lodges are determined.

#G5 [56g] In setting up the degrees of the circumference [of heaven], in each case take [the number of *du* indicated by] the marker corresponding to the leading [star of each lodge].

#G6 [59j] The position of the sun on rising and setting may also be determined by the [graduated] circumference. If you wish to know the rising and setting of the sun, then set out the 28 lodges using the 365 1/4 *du*.

#G7 [60b] When [the lodge] Well is centred at midnight [on the winter solstice], the beginning of Ox falls over the middle of *zi* 子 [due north]. When Well is 30 *du* and 7/16 *du* to the west of the central standard gnomon, and falls over the middle of *wei* 未 [30 degrees west of south], then the beginning of Ox falls over the middle of *chou* 丑 [30 degrees east of north]. With this, heaven and earth are matched together.

#G8 [60g] Now set out the circumference of the 28 lodges. When they are set out, set up a fixed gnomon at the centre of the graduated circumference. On the days of the winter and summer solstices, use this to sight on the sun as it first rises, and set up a displacement marker on the *du* so as to sight on the shadow of the central gnomon. When the shadow is produced [backwards], this indicates the degree of whatever lodge the sun is rising in, and similarly for the setting sun.

#G9 [60n] The distance of Ox from the pole is 115 *du* 1695 *li* 21 *bu* and 819/1461 *bu*.

[216] These are the stars marking the beginnings of the lodges named after the related asterisms.

#G10 [60p] Method:
Set up the distance from the outermost *heng* to the pivot of the north pole: 238 000 *li.*
Subtract 11 500 li for the *xuan ji*, leaving 226 500 *li*, which makes the dividend.
Take as divisor the size of one *du* on the innermost *heng*: 1954 *li* 247 *bu* and 933/1461 *bu.*
The integral quotient gives the number of *du.*
The *li* and *bu* are found from the remainder.
Take 1/300 of the rationalised remainder as the dividend, and take 1461 as the divisor, which gives the number of *li.*
Multiply the remainder by three, and division yields hundreds of *bu.*
Multiply the remainder by ten, and division yields tens of *bu.*
Multiply the remainder by ten again, and division yields units of *bu.*
Make the remainder the numerator of a fraction whose denominator is the divisor.
Proceed similarly in subsequent cases.

#G11 [62h] The distance of Harvester and Horn from the pole is 91 *du* 610 *li* 264 *bu* and 1296/1461 *bu.*

#G12 [62k] Method:
Set up the distance from the middle *heng* to the pivot of the north pole: 178 500 *li*, which makes the dividend.
Take as divisor the size of one *du* on the innermost *heng.*
The integral quotient gives the number of *du.*
The *li* and *bu* are found from the remainder.
Make the [final] remainder the numerator of a fraction whose denominator is the divisor.

#G13 [63b] The distance of Well from the pole is 66 *du* 1481 *li* 155 *bu* and 1245/1461 *bu.*

#G14 [63d] Method:
Set up the distance from the innermost *heng* to the pivot of the north pole: 119 000 *li.*
Add 11 500 *li* for the *xuan ji*, giving 130 500 *li,* which makes the dividend.
Take as divisor the size of one *du* on the innermost *heng.*
The integral quotient gives the number of *du.*
The *li* and *bu* are found from the remainder.
Make the [final] remainder the numerator of a fraction whose denominator is the divisor.

H. THE SHADOW TABLE

Introduction

This short section is clearly in good order. Its sole purpose is to provide a list of noon gnomon shadows for each of the 24 *qi* throughout the year. This is done by linear interpolation between the values for the summer and winter solstices, which are the only data to bear any close relation to observation. The results obtained are consistent with the calculations for the seven *heng* given in section #D.

In fact, as Zhao states explicitly, this shadow table is not part of the *Zhou bi* text seen by him. As he says, he was dissatisfied with the fact that the earlier table gave different shadow values for the spring and autumn equinoxes, and hence he set out to provide an improved version. Fortunately his description of the defects of the old table enable us to reconstruct it: see appendix 4. While the solstitial shadow values of the original table are certainly those associated with the *Zhou bi*, the failure to make the equinoctial shadow values equal make it difficult to see how this older document could be reconciled with the quantitative cosmography of the *Zhou bi* as represented in sections #B and #D.

Text

#H1 [63n] For the 24 *qi* 氣 of the 8 *jie* 節, the decrease or increase [of the shadow] for one *qi* is 9 *cun* 9 *fen* and 1/6 fen. The length of the winter solstice shadow is 1 *zhang* 3 *chi* 5 *cun*, and the length of the summer solstice shadow is 1 *chi* 6 *cun*. It is asked: What are the successive values of the varying lengths for each *jie*?

#H2 [63p][217]

	zhang	*chi*	*cun*	*fen*	*fen*/6
[1] winter solstice	1	3	5	0	0
[2] little cold	1	2	5	0	5
[3] great cold	1	1	5	1	4
[4] beginning of spring	1	0	5	2	3
[5] rain waters	0	9	5	3	2
[6] emerging insects	0	8	5	4	1
[7] spring equinox	0	7	5	5	0
[8] clear and bright	0	6	5	5	5

[217] This paragraph appears in tabular form in all editions. Numbering of the twenty-four *qi* has been added here for clarity, and units have been given at the heads of columns rather than being repeated after figures as in the original.

#H2 (continued)

	zhang	*chi*	*cun*	*fen*	*fen*/6
[9] grain rains	0	5	5	6	4
[10] beginning of summer	0	4	5	7	3
[11] grain fills	0	3	5	8	2
[12] grain in ear	0	2	5	9	1
[13] summer solstice	0	1	6	0	0
[14] little heat	0	2	5	9	1
[15] great heat	0	3	5	8	2
[16] beginning of autumn	0	4	5	7	3
[17] limit of heat	0	5	5	6	4
[18] white dew	0	6	5	5	5
[19] autumn equinox	0	7	5	5	0
[20] cold dew	0	8	5	4	1
[21] frost-fall	0	9	5	3	2
[22] beginning of winter	1	0	5	2	3
[23] little snow	1	1	5	1	4
[24] great snow	1	2	5	0	5

#H3 [65k] For the 24 *qi* of the 8 *jie*, the decrease or increase [of the shadow] for one *qi* is 9 *cun* 9 *fen* and 1/6 *fen*. The winter and summer solstices are the beginnings of decrease and increase.

#H4 [66a] Method:
Set up the winter solstice shadow, subtract the summer solstice shadow, and the difference is made the dividend.
Take 12 as the divisor.
The integral quotient gives the *cun*.
Multiply the remainder by ten, and divide again to obtain the *fen*.
Make the [final] remainder the numerator over the denominator.

I. LUNAR LAG

Introduction

This section shows no signs of corruption, and clearly stands by itself. It is concerned with the number of *du* by which the moon[218] is found to have shifted to the east of its former position after certain calendrically significant intervals. These intervals are:

[218] As throughout the *Zhou bi*, reference is to the mean moon.

(1) The day.
(2) The 'small year' of twelve mean lunations.
(3) The 'large year', including an intercalary month.
(4) The 'mean year', the average length over the 19-year cycle *zhang*章.
(5) The 'small month' of 29 days.
(6) The 'large month' of 30 days.
(7) The mean lunation.

In each case the result is stated, and is then followed by step by step instructions for performing the relevant computation.

The only link between this section and the rest of the *Zhou bi* is its use of the basic constants of the quarter-remainder system to track the moon's angular motion. As always, this is (in our terms) the mean moon rather than the real one with its complex variations of speed. In its use of the term *she* 舍 rather than *xiu* 宿 for the lunar lodges, this section differs from the practice of sections #B and #G. Clearly anybody wishing to know the moon's position at some future epoch would find the results given here some use in avoiding counting board operations involving inconveniently large numbers. Starting from the moon's present known position, one would simply total up the amount of lag attributable to the years, months and days elapsed. A wide range of astronomical/hemerological procedures could thereby be performed with greater facility. Compare the treatment in *Han shu* 21b, which deals with the similar problems by the methods of the *San tong li*.

Text

#I1 [67g] The moon lags behind heaven by 13 *du* and 7/19 *du* [per day].

#I2 [67j] Method:
Set up the months in a *zhang* 章: 235.
Divide by the years in a *zhang*: 19
Add the [daily] solar motion: 1 *du*.
The result is 13 *du* and 7/19 *du*.
This is the size of the moon's daily motion, which is the amount of *du* it lags heaven.

#I3 [67m] In a small year the moon lags behind its old lodge by 354 *du* and 6612/17 860 *du*.

#I4 [68b] Method:
Set up a small year: 354 days and 348/940 days.
Multiply by the moon's lag behind heaven: 13 *du* and 7/19 *du*.
Take this as the dividend.
Take as divisor the product of the degree denominator and the day denominator.
On dividing, the result for the accumulated lag behind heaven is 4737 *du* and 6612/17 860 *du*.
Divide this by the circumference of heaven: 365 *du* and 4465/17 860 *du*.
The remainder is 354 *du* and 6612/17 860 *du*.
This is the number of *du* by which the moon falls short of its old lodge. All other cases may be treated similarly.

#I5 [69g] In a large year the moon falls short of its old lodge by 18 *du* and 11 628/17 860 *du*.

#I5 [69j] Method:
Set up a large year: 383 days and 847/940 days.
Multiply by the moon's lag behind heaven: 13 *du* and 7/19 *du*.
Take this as the dividend.
Take as divisor the product of the degree denominator and the day denominator.
On dividing, the result for the accumulated lag behind heaven is 5132 *du* and 2698/17 860 *du*.
Divide this by the circumference of heaven.
The remainder is the number of *du* by which the moon falls short of its old lodge.

#I6 [70g] In a mean year the moon falls short of its old lodge by 134 *du* and 10 105/17 860 *du*.

#I7 [70j] Method:
Set up a mean year: 365 days and 235/940 days.
Multiply by the moon's lag behind heaven: 13 *du* and 7/19 *du*.
Take this as the dividend.
Take as divisor the product of the degree denominator and the day denominator.
On dividing, the result for the accumulated lag behind heaven is 4882 *du* and 14 570/17 860 *du*.
Divide this by the circumference of heaven.
The remainder is the number of *du* by which the moon falls short of its old lodge.

#I8 [71d] In a small month the moon falls short of its old lodge by 22 *du* and 7755/17 860 *du*.

#I9 [71f] Method:
Set up a small month 29 days.
Multiply by the moon's lag behind heaven: 13 *du* and 7/19 *du*.
Take this as the dividend.
Take as divisor the product of the degree denominator and the day denominator.
On dividing, the result for the accumulated lag behind heaven is 387 *du* and 12 220/17 860 *du*.
Divide this by the circumference of heaven.
The remainder is the number of *du* by which the moon falls short of its old lodge.

#I10 [72a] In a large month the moon falls short of its old lodge by 35 *du* and 14 335/17 860 *du*.

#I11 [72c] Method:
Set up a large month 30 days.
Multiply by the moon's lag behind heaven: 13 *du* and 7/19 *du*.
Take this as the dividend.
Take as divisor the product of the degree denominator and the day denominator.
On dividing, the result for the accumulated lag behind heaven is 401 *du* and 940/17 860 *du*.
Divide this by the circumference of heaven.
The remainder is the number of *du* by which the moon falls short of its old lodge.

#I12 [72m] In a mean month the moon falls short of its old lodge by 29 *du* and 9481/17 860 *du*.

#I13 [71f] Method:
Set up a mean month: 29 days and 499/940 days.
Multiply by the moon's lag behind heaven: 13 *du* and 7/19 *du*.
Take this as the dividend.
Take as divisor the product of the degree denominator and the day denominator.
On dividing, the result for the accumulated lag behind heaven is 394 *du* and 13 946/17 860 *du*.
Divide this by the circumference of heaven.
The remainder is the number of *du* by which the moon falls short of its old lodge.

J. RISING, SETTING AND SEASONS

Introduction

This section has little evident relevance to others in the *Zhou bi*, and the reasons for its inclusion are at first sight unclear. In the first two paragraphs the sun's rising and setting at the solstices are referred to the twelve-fold azimuthal division of the horizon labelled by the cyclical characters of the *di zhi* 地支'earthly branches'. In #J4 and #J5 rising and setting are referred to a parallel eightfold division of the horizon labelled by the *ba gua* 八卦 eight trigrams. The two systems are shown in figure 15. From the way in which the seasonal risings and settings are said to be related to the trigrams, it is clear that they have been arranged in the so-called 'King Wen 文' order. The data given for each system are approximately true for latitudes near 35 degrees north, which is the region in which the solstitial shadows of the *Zhou bi* appear to have been measured. The material is gnomic rather than expository, and perhaps therefore is bound to seem rather fragmentary. There does not seem to be any obvious way of re-arranging the paragraphs to read more smoothly.

The data given here cannot be related closely to the observational or computational procedures given in the rest of the work. On the other hand, we have already seen one instance in #G7 where the graduations borne by the *shi* 式 cosmic model were clearly present in the consciousness of a mainstream *Zhou bi* theorist. Since the earth-plate of the *shi* bears both the twelve stems and the eight trigrams, it may be that the role of the *shi* as a physical representation of the *gai tian* cosmography may be the missing link between this section and the rest of the book. Zhao points out the resemblance between #J7 and a passage from the Book of Change, an appropriate quotation if the picture of a divination aid is in the author's mind.

Text

#J1 [73j] At the winter solstice the day is at its shortest. The sun rises at *chen* 辰 and sets at *shen* 申. The Yang illuminates three parts, but does not cover nine. The east and west points are opposite each other due south.

#J2 [73l] At the summer solstice the day is at its longest. The sun rises at *yin* 寅 at sets at *xu* 戌. The Yang illuminates nine parts, but does not cover three. The east and west points are opposite each other due north.

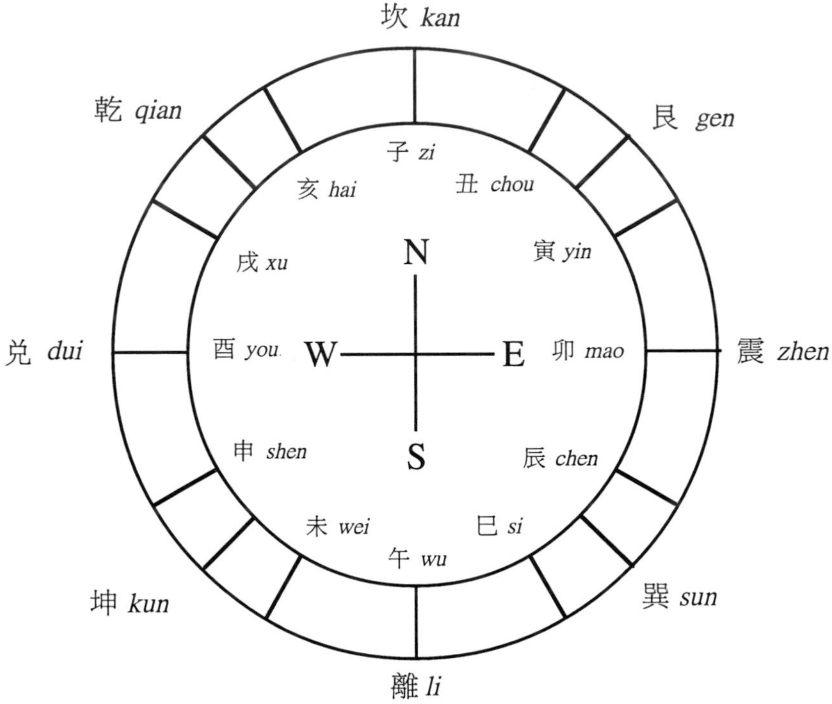

Figure 15. Horizon divisions.

#J3 [74d] The sun rises on the left and sets on the right; it moves north and south.

#J4 [74e] Therefore the winter solstice follows [the trigram] *kan* 坎. The Yang is at *zi* 子. The sun rises at *sun* 巽 and sets at *kun* 坤. The sun's rays are seen to be scanty; thus it is said to be cold.

#J5 [74f] The summer solstice follows *li* 離. The Yin is at *wu* 午. The sun rises at *gen* 艮 and sets at *qian* 乾. The sun's rays are seen to be plentiful; thus it is said to be hot.

#J6 [74h] If the sun and moon miss their degrees, cold and heat will become confused with each other.

#J7 [74l] That which departs is contracted; that which comes is extended. Therefore the contracted and extended respond to each other.

#J8 [74n] Therefore, after the winter solstice the sun moves to the right. After the summer solstice the sun moves to the left. That which moves to the left is departing, and that which moves to the right is coming.[219]

K. CALENDRICAL CYCLES

Introduction

In this final section we return to calendrical topics with a discussion of the basic constants of astronomical systems of the quarter-remainder type. Apart from the fact that #K6 is almost certainly a fragment of intruded comment, it is possible to read the section as it stands without much sense of confusion.

After a brief definition of the month, the day and the year as the basic cycles of the calendar, and a reference to the sequence of *qi* 氣 (#K1 to #K3), the text names five longer period cycles (#K4). The first four are well-known in the context of the quarter-remainder system type, and serve computational convenience in different ways. The fifth cycle is however based on a multiple of seven, a number which has no place in the structure of traditional Chinese time reckoning. To my knowledge a cycle of this kind has no parallels elsewhere. An inevitable speculation is that the factor of seven may indicate some awareness of the existence of the seven-day week. For further discussion see page 25.

Paragraph #K5 asks the five questions set out in table 5, whose answers take up the rest of the section.

Table 5. Calendrical queries

QUERY	RESPONSE
How is it known that [the circumference of]	
heaven is 365 1/4 *du*,	#K8 to #K11
that the sun moves one *du* [in a day],	[#K8 by implication]
and that the moon lags heaven by 13 *du* and 7/19 *du*,	#K14
that 29 days and 499/940 days make one month,	#K16
and that 12 months and 7/19 months make one year?	#K15

[219] We are probably better off thinking in terms of the correlations left:right :: yang:yin :: departing:coming than trying to think what precise orientation of an observer to what particular phenomena can be referred to here.

The fact that this section concludes as soon as these questions have received satisfactory answers suggests that its final few paragraphs are undamaged, and hence perhaps that the book as a whole has not lost material from its end.

Text

#K1 [75a] Therefore[220] a conjunction of the moon with the sun makes one month. When the sun returns to [the previous position of] the sun [in the sky], that makes one day. When the sun returns to a star, that makes one year.[221]

#K2 [75b] The outermost *heng*: winter solstice. The innermost *heng*: summer solstice.

#K3 [75b] The six *qi* return in sequence, and are all called 'medial *qi*'.

#K4 [75e] The reckoning of Yin and Yang, the methods for days and months:
19 years make a *zhang* 章.
4 *zhang* make a *bu* 蔀, 76 years.
20 *bu* make a *sui* 遂, and a *sui* is 1520 years.
3 *sui* make a *shou* 首, and a *shou* is 4560 years.
7 *shou* make a *ji* 極, and a *ji* is 31 920 years.
All the reckonings of generation come to an end, and the myriad creatures return to their origin. [From this new origin,] Heaven creates the chronological reckoning [once more].

#K5 [77a] How is it known that [the circumference of] heaven is 365 1/4 *du*, that the sun moves one *du* [in a day], and that the moon lags heaven by 13 *du* and 7/19 *du*, that 29 days and 499/940 days make one month, and that 12 months and 7/19 months make one year?

#K6 [78b] Divide [by] the circumference of heaven. As for the remainder, make it the numerator over the conjunction [denominator].

#K7 [78c] In ancient times, Bao Xi 包犧 and Shen Nong 神農 created the calendar, and planned the beginning of the origin [or perhaps 'epoch' here, taking the more technical sense of *yuan* 元]. They saw that the three luminaries were not yet in accord

[220] *Gu* 故, which I translate here conventionally as 'therefore' is clearly performing its not uncommon function as a kind of paragraph marker rather than as a serious logical connective.

[221] No distinction was made in Han times between the tropical year and the sidereal year.

with the proper pattern, and that the sun, moon and stars as yet had no allotted degrees.

#K8 [78d] The sun rules the daylight, and the moon rules the night: daylight and night make one day. The sun and moon start off together from the Establishment star (*Jian xing* 建星).[222] The moon's degrees are rapid, and the sun's degrees are slow. The sun and moon meet again in between 29 and 30 days, when the sun has moved a little over 29 *du* round heaven, the amount of this fraction not being fixed.

#K9 [78g] So after 365 [days] the sun is [once more] at its southernmost point and the shadow is long[est]. The next day it shortens again, and at the end of a year the solar shadow is at its long[est] again.

#K10 [78h] Therefore we know that in three periods of 365 days and in one period of 366 days [the solstice occurs at the same time of day as before].

#K11 [78j] Therefore we know that one year is 365 1/4 days; 'year' (*sui* 歲) means 'conclusion' (*zhong* 終).

#K12 [78k] The moon has an accumulated lag behind heaven amounting to 13 revolutions, plus a little over 134 *du*, ignoring [the fact that] the [daily] lag behind heaven of 13 *du* and 7/19 *du* is not constant.

#K13 [78m] So, [when] the sun has made 76 circuits, the moon has made 1016 circuits, and they reach conjunction [once more] at the Establishment star.

#K14 [79a] Set the amount by which the moon's motion lags heaven, and divide by the amount by which the sun lags heaven. The result is 13 *du* and 7/19 *du*, which is the daily motion of the moon over heaven.

#K15 [79c]

Again, set up the accumulated months in 76 years, and divide by 76 years. We obtain 12 months and 7/19 months, which is the number of months in a year.

#K16 [79e] Set up the number of degrees in the circumference of heaven, and divide it by 12 7/19 months. We obtain 29 499/940 days, which is the number of days in a month.

[222] This star, nowadays identified as π Sgr, is evidently a marker for the winter solstice, the point at which the sun and moon are conventionally in conjunction at the moment of system origin from which calendrical cycles are reckoned. Other sections of the *Zhou bi* use the determinative star of the lodge Ox.

Appendix One

Zhao Shuang and Pythagoras' theorem

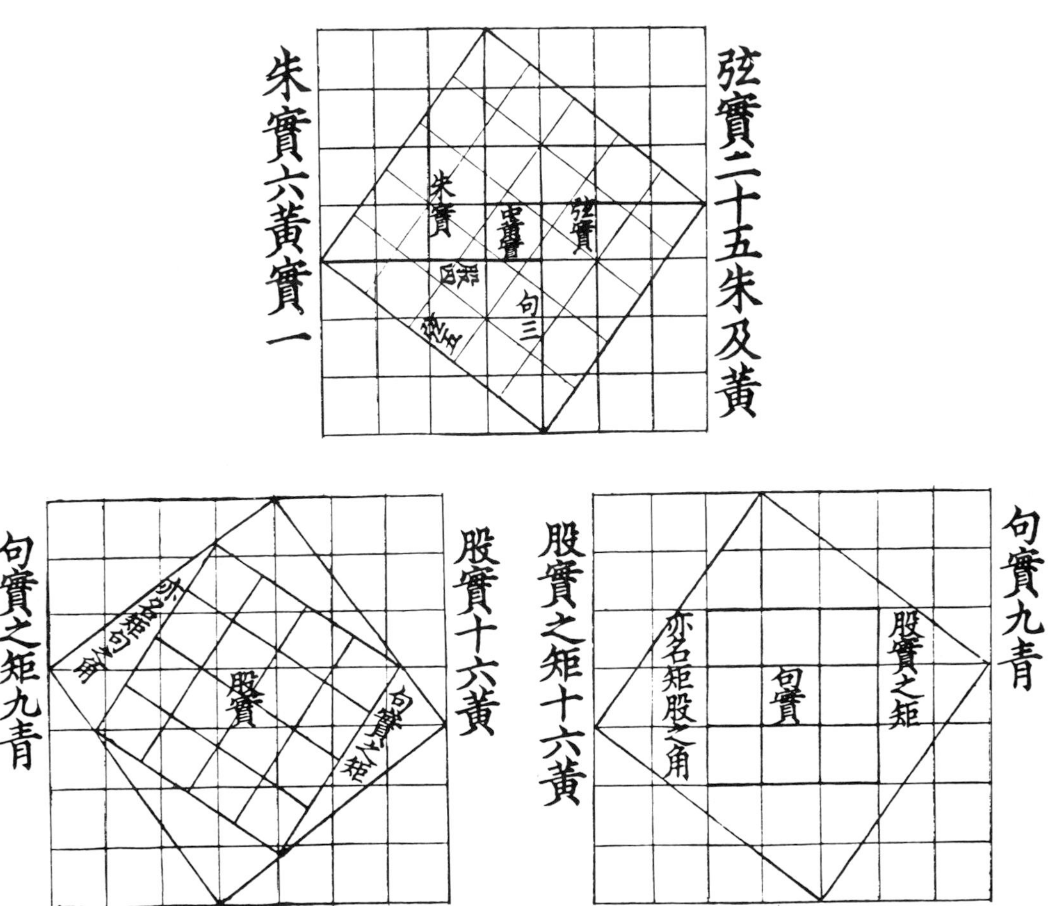

Figure 16. Hypotenuse, left and right diagrams from Song edition.

Perhaps in an attempt to compensate for his evident failure to make real sense of the Pythagorean material in section #A, Zhao adds a short essay which discusses various general mathematical relations between base, altitude and hypotenuse, and the areas of plane figures that can be formed by combining them. This is found between paragraphs #A3 and #A4 in all editions: see p.18 in Qian's edition. His essay is followed by comments by Zhen Luan, who attempts to illustrate Zhao's statements by using the example base 3, altitude 4 and hypotenuse 5. Zhen's comments are in turn the subject of comment by Li Chunfeng, who corrects him at several points.

In all traditional editions this essay is prefaced by the three diagrams shown in figure 16: hypotenuse, right and left. It is clear, however, that the second and third of these do not relate at all to Zhao's essay, except perhaps as desperate but mistaken attempts to penetrate its meaning. After having studied Zhao's essay in detail, I think it is clear that the two explicit mentions of a 'diagram' in his text both refer to the first of these, the 'hypotenuse diagram' *xian tu* 弦圖 (figure 17). This is equivalent to the final diagram restored by Qian, and includes the first of Qian's diagrams within itself. The 'right' and 'left' diagrams look like an effort to supply illustrations to cover the cases illustrated by me in figures 23 and 19, by someone who could see that a 3-side and 4-side square had to be somehow got inside the 5-side square, but did not realise what Zhao meant by the expression 'trysquare'. It might then seem natural to place these inner squares at an angle relative to the immediately circumscribing 5-side square, just as that square was rotated relative to its circumscribing 7-side square. It is possible that Zhen Luan supplied these diagrams himself. The insertion (if that is the right explanation) seems to have happened by the time of Zhen Luan in the sixth century, since he refers to the diagrams in his commentary. Thus we find Zhen using the following expressions:

18s: 'the central yellow area on the left diagram'. Why Zhen should refer to the left diagram when Zhao has just directed attention to the hypotenuse diagram is a mystery. Furthermore, Zhen misrecites Zhao's 'the difference of base and altitude' a few characters earlier as 'the difference of base and hypotenuse' with disastrous results. Perhaps the text seen by Zhen had *no* hypotenuse diagram, so that he did not understand what was meant? And perhaps the text he saw was indeed corrupted, but Li Chunfeng used a better version (which also had the hypotenuse diagram) in preparing the text which we now see, while copying in Zhen's comments uncorrected?

19j: 'the outer green [area] on the left diagram'. While the area is 25 – 16 = 9 as required, the trysquare shape is completely lost.

19j 'the central yellow [area] of 16 on the left diagram'. The area and shape are correct, but the square is misplaced.

20b 'Take double the altitude of 4, one gets 8; make this the *congfa* 從法 on both sides of the diagram, and open the corner 9 of the trysquare base, getting 1'. Zhen does not specify which diagram he is talking about, but he is clearly doing no more than shuffling numbers since he does not understand the situation Zhao is discussing – see figure 20 below.

21f–21m: Although Zhen speaks here as if he understood the hypotenuse diagram, he may simply be following Zhao's text, which is explicit and clear enough for him to do this.

22a–22l: Zhen clearly does not visualise the situation described by Zhao: see figure 28. He is corrected by Li Chunfeng.

This essay by Zhao was the subject of a detailed and ingenious study by Brendan S. Gillon, in which independently of Qian he discussed the problem of restoring the corrupted diagrams, partly on the basis of Liu Hui's commentary to the *Jiu zhang suan shu* (AD 263).[223] With the benefit of hindsight, I suspect that some of Zhao's essay may be rather simpler than Gillon's rendering may suggest, and I have tried to reflect that in my translation. In what follows I limit myself as far as possible to a straightforward rendering of Zhao's text, which is not too difficult to comprehend so long as one is looking at the right diagram. I have therefore supplied these with explanatory captions.[224] So far as we can tell, Zhao expected his readers to be able to do this for themselves, apart from the case of the 'hypotenuse diagram' which he mentions explicitly.

Text

Diagrams of base and altitude, circle and square[225]

The base and altitude are each multiplied by themselves. Add to make the hypotenuse area. Divide this to open the square,[226] and this is the hypotenuse.

In accordance with the hypotenuse diagram [figure 17], you may further multiply the base and altitude together to make two of the red areas. Double this to make four of the red areas. Multiply the difference of the base and the altitude by itself to make the central yellow area. If one [such] difference area is added [to the four red areas], the hypotenuse area is completed.

[223] Gillon (1977), 253–93.

[224] It seems to me important to avoid algebraic manipulation in explaining the processes referred to by Zhao, except perhaps where it is simpler to write $(a + b)$ rather than 'the sum of a and b'. Zhao had no algebra, and so far as the content of this essay goes seems to have understood his mathematics solely in terms of operations involving geometric magnitudes.

[225] As Zhao makes clear in his commentary to #A2, the reason that circle and square are mentioned here is that the perimeter of a circle of unit diameter is taken as three, while the perimeter of a unit sided square is four. Three and four are of course the lengths of base and altitude considered in #A. This answers the problem raised in Gillon (1977), 289.

[226] This is the full Chinese expression for root extraction, which was accomplished by a process similar to division: see Martzloff (1988), 210ff. Later on Zhao just tells us to 'open' an area.

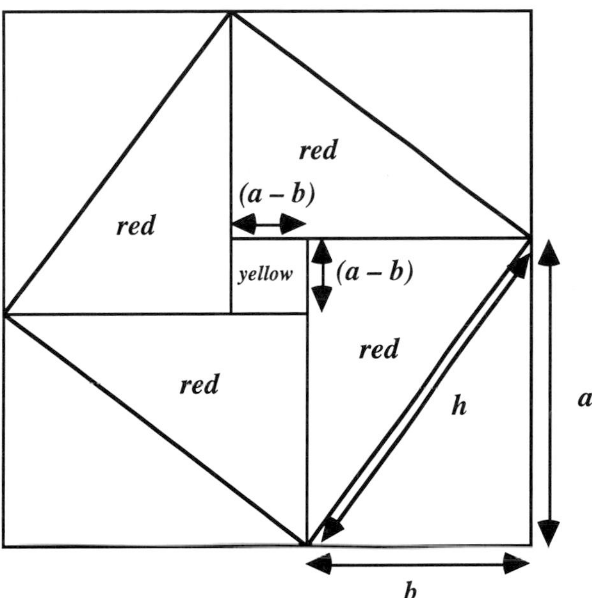

Figure 17. Hypotenuse diagram.

Subtract the difference area from the hypotenuse area, and halve the excess. Take the difference as the 'auxiliary divisor'.[227] Divide to open the square, and once more you obtain the base. [See figure 18.] If you add the difference to the base, you get the altitude.Whenever you add the base and altitude areas you form the hypotenuse area. [Either of these] may be a square within, or a trysquare without.[228] The shapes are transformed[229] but the quantities are balanced; the forms are different but the numbers are equal.

[227] This is found elsewhere as a technical term in the solution of what we would nowadays call second degree equations with a linear term: see Li and Du (1987), 54. Gillon rightly points out that in his commentary on the *Jiu zhang suan shu* 九章算術 Liu Hui refers to diagrams like the ones we are using here when he explains how to find solutions to such problems. There are however no signs that Zhao is thinking of solving problems of this kind, *pace* Li and Du, 63–5.

[228] I follow Qian in this reading. Qian notes (p. 19 collation note 2) that extant texts have 'square' and 'trysquare' in the reverse order, which does not accord well with Zhao's later statements, represented in figs. 19 and 23, in which the squares are said to be *fang qi li* 方其裡 'squaring the inside' of the relevant trysquare. Zhao is simply explaining the general principle behind much of his subsequent discussion.

[229] The word used here is *gui* 詭, with a number of senses connected with deception or peculiarity. Here I follow the sense of a similar passage in Liu Hui's commentary to the fifth problem of chapter 9 of the *Jiu zhang suan shu*, which has the more comprehensible *e* 訛 'change, transform' both here and in Zhen Luan's recital of this passage in his commentary. The two characters are graphically similar, and it is possible that Zhen (or an intervening copyist) simply misread Zhao's text.

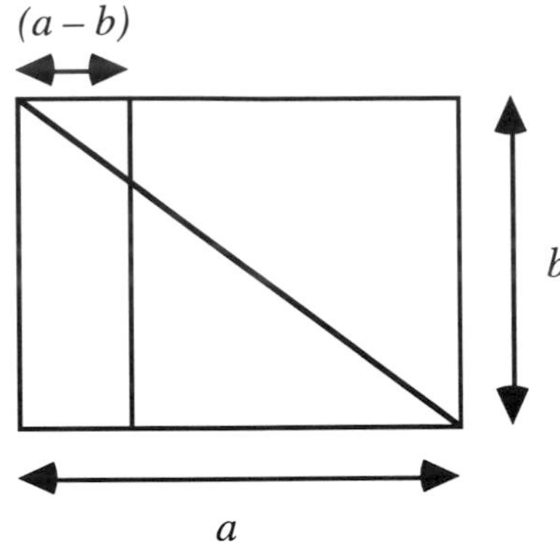

Figure 18. $(a - b)$ as auxiliary divisor.

The trysquare of the base area has the altitude–hypotenuse difference as its breadth, and the altitude–hypotenuse sum as its length, while the altitude area is the square within. [See figure 19.] Subtract the area of the trysquare base from the hypotenuse area, 'open' the excess and it is the altitude. [Make] twofold [use of] the altitude on both sides as the 'auxiliary divisor',[230] open the corner of the trysquare base [which now remains] and it is the altitude–hypotenuse difference. [See figure 20.] Add [it] to the altitude to make the hypotenuse.

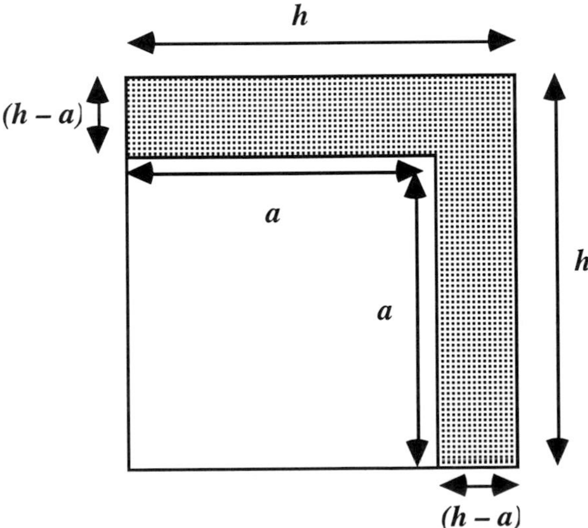

Figure 19. Base area trysquare.

[230] Here this seems to mean little more than that we lay this length off along two sides.

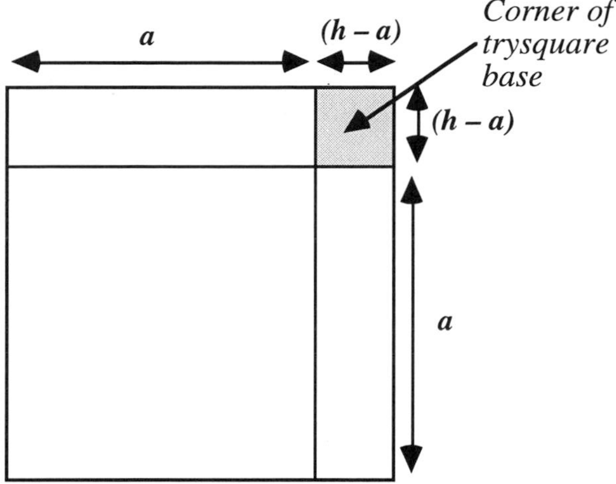

Figure 20. Corner of trysquare base.

Divide the base area by [this] difference, and you obtain the altitude–hypotenuse sum. Divide the base area by [this] sum, and thus you obtain the altitude–hypotenuse difference.

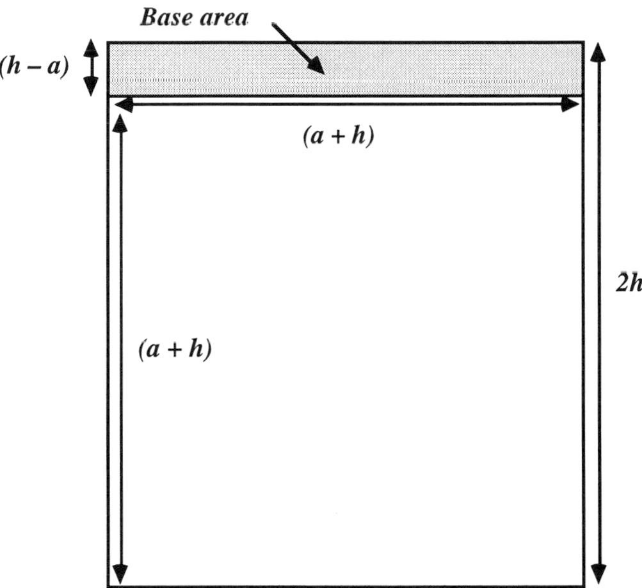

Figure 21. Hypotenuse from sum and difference of altitude and hypotenuse (not to scale).

'Let the sum [of altitude and hypotenuse] be multiplied by itself, and add it to the base area to create a [new] area. Take double the sum as the divisor, and what you obtain is then the hypotenuse. [See figure 21.] Subtract the base area from the sum multiplied by itself, [and] divide [by double the sum as before] to make the altitude.[See figure 22.]

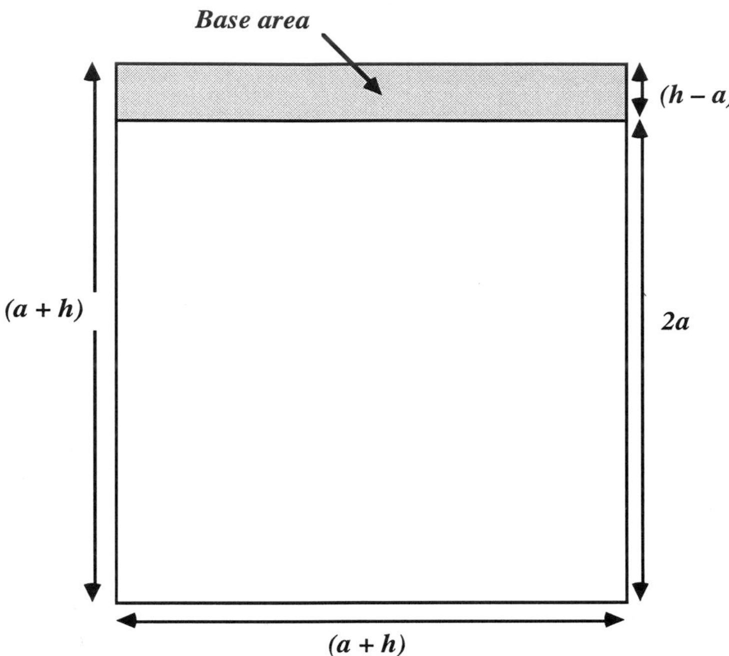

Figure 22. Altitude from sum and difference of altitude and hypotenuse (not to scale).

The trysquare of the altitude area has the base–hypotenuse difference as its breadth, and the base–hypotenuse sum as its length, while the base area is the square within. [See figure 23.] Subtract the area of the trysquare altitude from the hypotenuse area, 'open' the excess and it is the base. [Make] twofold [use of] the base on both sides as the 'auxiliary divisor', 'open' the corner of the trysquare altitude [which now remains] and it is the base–hypotenuse difference. [See figure 24.] Add [it] to the base to make the hypotenuse.

Divide the altitude area by [this] difference, and you obtain the base–hypotenuse sum. Divide the altitude area by [this] sum, and thus you obtain the base–hypotenuse difference.

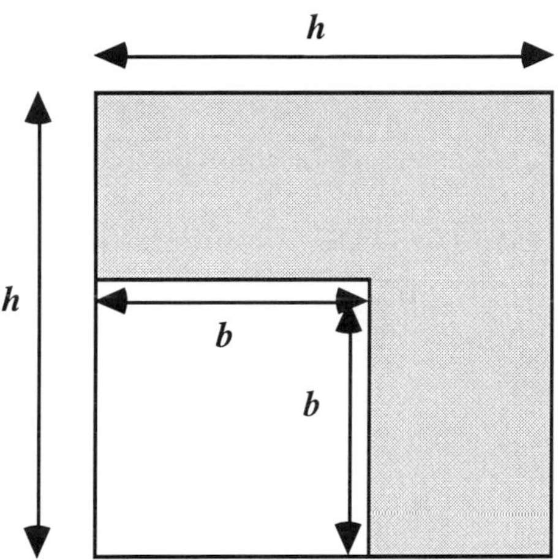

Figure 23. Altitude area trysquare.

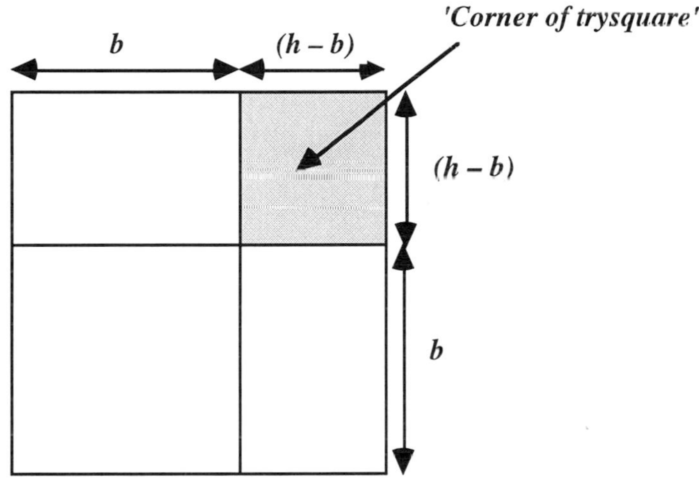

Figure 24. Corner of trysquare altitude.

Let the sum [of base and hypotenuse] be multiplied by itself, and add it to the altitude area to create a [new] area. Take double the sum as the divisor, and what you obtain is then the hypotenuse.[See figure 25.] Subtract the altitude area from the sum multiplied by itself, [and] divide [by double the sum as before] to make the base. [See figure 26.]

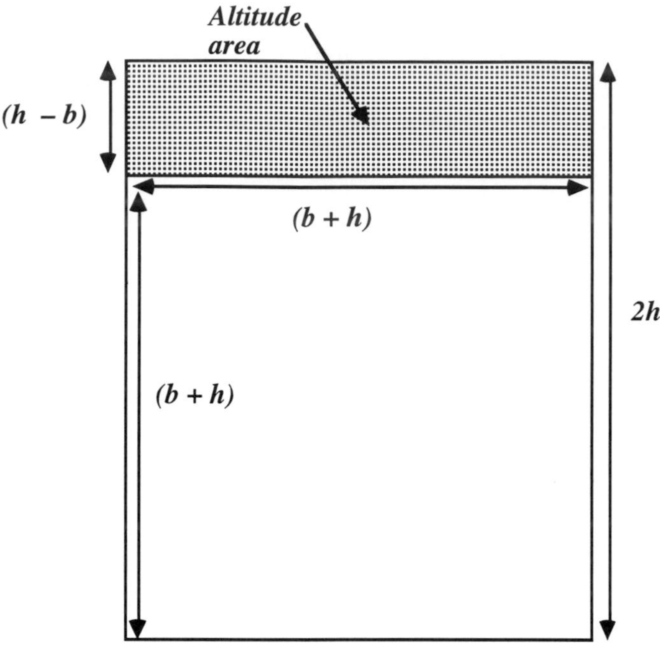

Figure 25. Hypotenuse from sum and difference of base and hypotenuse (not to scale).

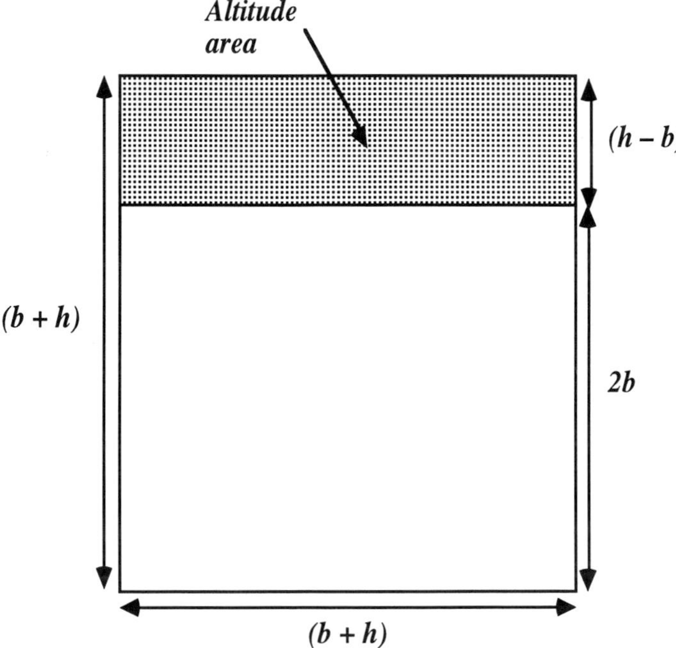

Figure 26. Base from sum and difference of base and hypotenuse (not to scale).

Multiply the two differences together, double [the resulting area] and 'open' it. [As for] what is obtained, [if] you increase it by the altitude–hypotenuse difference, you make the base. [If] you increase it by the base–hypotenuse difference, you make the altitude. [If] you increase it [by] both differences, you make the hypotenuse. [See figure 27.]

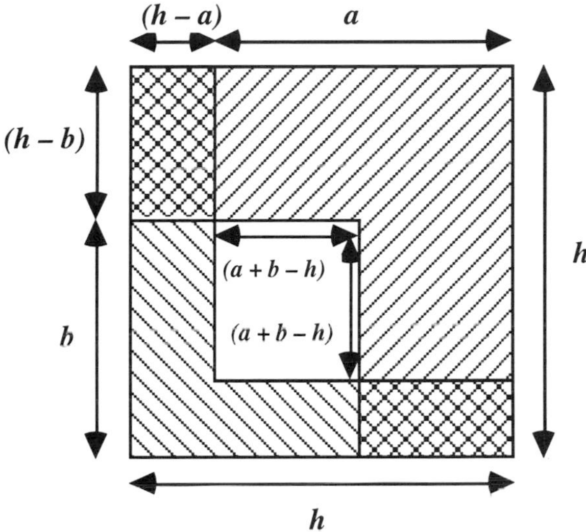

Figure 27. To obtain the three sides from $(h-a)$ and $(h-b)$, perhaps showing the base and altitude trysquares 'plaited round' 盤桓 *pan huan* as described in #A3.

As for the fact that you see the 'sum area' [i.e. the square of the sum of base and altitude] when you double the hypotenuse area and set out the base–altitude difference area [in addition to it], examining [the matter] by means of the diagram,[231] doubling the hypotenuse area fills the outer large square, [leaving as] excess the yellow area. Now the excess of this yellow area is just the base–altitude difference area. If you take away the difference area [from double the hypotenuse area], and 'open' the excess, you get the outer large square,[232] [and then] the side of the outer large square, which is the base–altitude sum. Let this sum be multiplied by itself, [take] double the hypotenuse

[231] Zhao is now referring to the whole of the traditional 'hypotenuse diagram', the last of Qian's restored diagrams.

[232] Although the sentence is not nonsense as it stands, it would make better sense and conform to the style of the rest of the essay better if this first occurrence of the words 'the outer large square' were excised. After an instruction to perform an arithmetical operation, the usual Chinese practice is to tell you what you get as a result of the operation, not as here to tell you what you have before and after it. On the other hand the subsequent reference to 'the central yellow square' follows the same pattern, so perhaps the text is as Zhao wrote it.

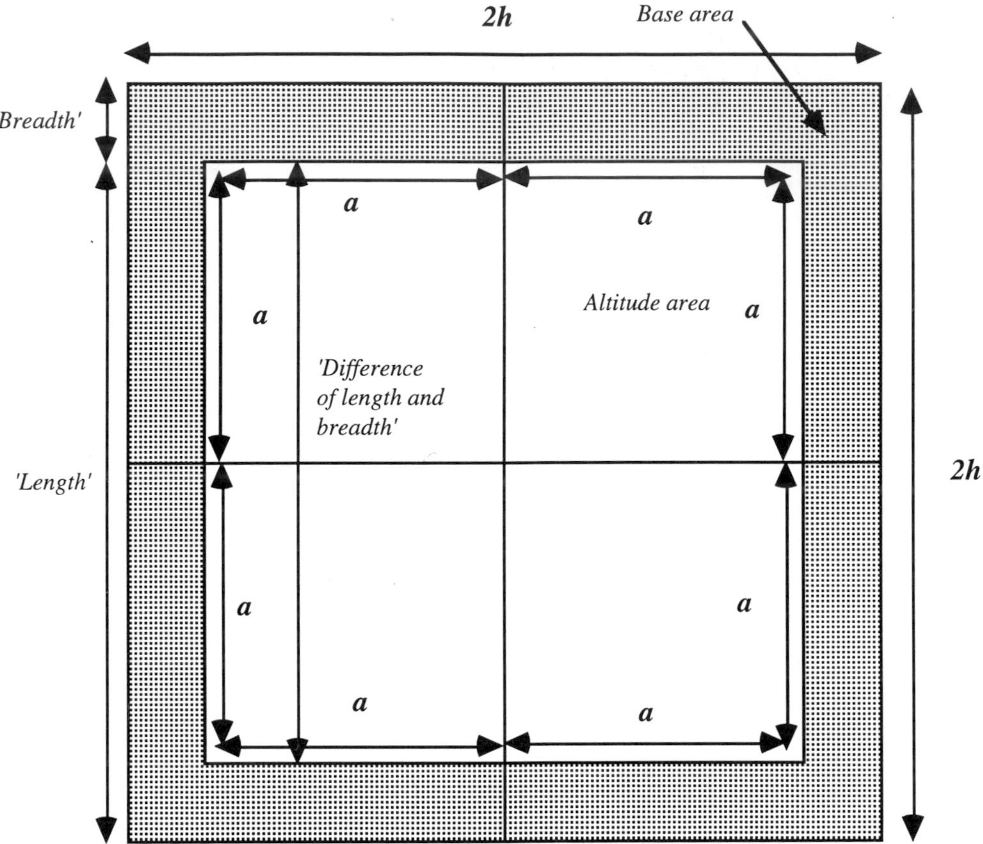

Figure 28. Using sum and difference of length and breadth of a trysquare (in this case the base trysquare is shown).

area then subtract [the result] from it, 'open' the excess, and you obtain the central yellow square. The side of this central yellow square is the base–altitude difference. Subtract this difference from the [base–altitude] sum and halve, to make the base. Add the difference to the sum and halve, to make the altitude.

Let the hypotenuse be [used] twice to make the joining of length and breadth [of the trysquares referred to earlier]. Let whichever of base or altitude appears [as its 'trysquare' in the case considered] be multiplied by itself to give an area. [Take] four [of these] areas and subtract [them] from it [i.e. the square whose side is double the hypotenuse]. 'Open' the excess, and what you get is the difference [between length and breadth]. Subtract this difference from the joining [of length and breadth], halve the surplus and you have the breadth. Subtract the breadth from the hypotenuse, and you have what you seek [i.e. the so far unknown side]. [See figure 28.]

If you observe the alternation of compasses and trysquare, how they return and repeat [their operations] together, how they mutually penetrate [but have] an allotted [function], there is something to be gained from each [instance]. Thus they systematise the multitude of rules, and vastly order the manifold principles, threading through the obscure and entering into the subtle, hooking up the profound and reaching afar. So [as the text of the *Zhou bi* says], what they do is simply to settle and regulate everything there is.[233]

[233] See #A7; Zhao modifies the quotation slightly.

Appendix Two

Zhao Shuang and the height of the sun

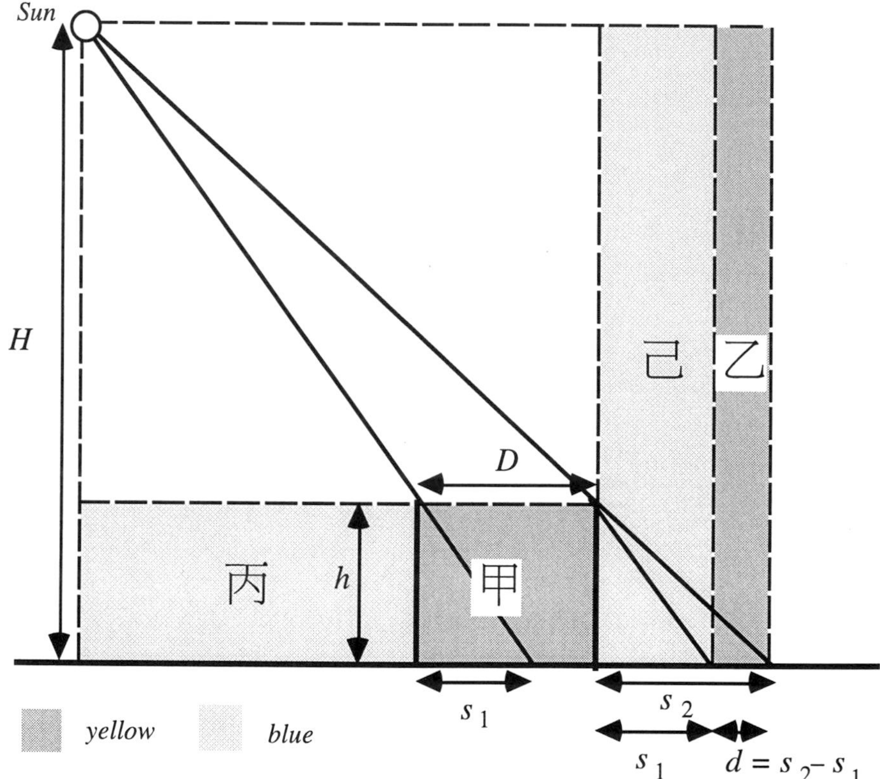

Figure 29. Restoration of Zhao's diagram, showing two gnomons of height h a distance D apart, casting shadows of length s_1 and s_2, with the sun at height H above the earth.

In #B11 Chen Zi tells Rong Fang that the sun, and hence the sky, is 80 000 *li* above the earth. After this paragraph Zhao Shuang inserts a short essay on the use of the gnomon to determine the sun's position. This is found between #B11 and #B12: see p. 34 in Qian's edition. Zhao' discussion refers explicitly to a diagram, which some editions have preserved although in garbled form: see the restored version of figure 29. Zhao's text runs as follows:

> The areas [marked with] a yellow *jia* 甲 and with a yellow *yi* 乙 are exactly equal. The area [marked with] a yellow *jia* is made by multiplying the height of a gnomon by the distance between them. If you count one for each time [you subtract] the shadow difference [i.e. divide by the shadow difference], which is the breadth of the [area marked with] a yellow *yi*, what you obtain is transformed into the length of the [area marked with] a yellow *yi*, which goes up level with the sun. Referring to the diagram, [it is evident that] you must add the height of the gnomon. Now when [the text] says 80 000 *li*, this is to be added on top of the gnomon.
>
> The [areas marked with] a blue *bing* 丙 and a blue *ji* 己 are likewise equal. If you amalgamate the [areas marked with] a yellow *jia* and a blue *bing*, and amalgamate the [areas marked with] a yellow *yi* and a blue *ji*, the [resulting] areas are likewise equal.

The relations stated by Zhao are true, but do not seem immediately obvious from the diagram. There is however no need for assuming any elaborate mathematics behind Zhao's claims. Consider for instance his claim that the yellow *jia* area must equal the yellow *yi* area. For the first, we have:

$$\text{area} = (\text{distance between gnomons}, D) \times (\text{height of gnomon}, h).$$

While for the second we have:

$$\text{area} = (\text{difference in shadows}, d) \times (\text{height of sun}, H).$$

Now in #B11 Chen Zi has just made it clear that a measurement at the gnomon can be transformed into the related large-scale measurement by changing each *cun* to 1000 *li*. The two first quantities on the right hand sides are related in this way, as are the two second quantities, but inversely. The products must therefore be the same. Zhao gives us little help in deciding what areas he marked with the blue characters *bing* 丙 and *ji* 己. Those shown here do however seem plausible candidates, and they obey the relations given by Zhao, in a similar fashion to the yellow areas.

Once again it is clear that a Chinese mathematician of this era did not think in terms identical to our own. Whereas we would automatically use similar triangles in such a situation, Zhao reaches an equivalent result by different means. This short essay is paralleled by a suggestion by Liu Hui in his preface to the *Jiu zhang suan zhu*, possibly composed close to the time that Zhao wrote. Apparently dissatisfied with calculations based on the old shadow rule, he proposed to use the *chongcha* 重差 'repeated difference' and *gougu* 勾股 'base and altitude' methods to attack the problem afresh:

> Set up two gnomons at the city of Yangcheng 陽城.[234] Let them be eight *chi* tall, and to north and south let the ground be quite level. On the same day, precisely at noon, take the shadow difference as the divisor, multiply the gnomon height and the distance between the gnomons to make an area, count one for each [subtraction] of the divisor [i.e. divide and take the integral result], add what you get to the gnomon height, and it is the distance of the sun from the earth.
>
> Multiply the shadow length at the southern gnomon by the distance between the gnomons to make an area, count one for each [subtraction] of the divisor, and you get [the distance] from the southern gnomon to the subsolar point.
>
> Take the distance to the subsolar point and the distance of the sun from the earth as base and altitude, and by means of these find the hypotenuse, which is the distance of the sun from the observer.
>
> Use a tube of [internal] diameter one *cun* to sight southwards on the sun [and adjust its length so that] the sun just fills the bore of the tube. Then the length of the tube gives the proportion for the altitude, and the diameter gives the proportion for the base. Now the amount of the sun's distance from the observer is the 'great altitude' and the base corresponding to it [in proportion as established] is the diameter of the sun.

Liu's work is not identical to Zhao's, but the correspondence is quite close. The reference to the sighting tube is a clear parallel with paragraph #B11 of the *Zhou bi* itself. There is no doubt that contact of some kind has occurred, even if we cannot be sure of its precise means or nature. Other evidence suggests that the more likely possibility is that Zhao is drawing on Liu rather than the other way around: see page 160f.

234 In Henan: this was the traditional site for gnomon observations.

Appendix Three

Zhao Shuang and the diagram of the seven *heng*

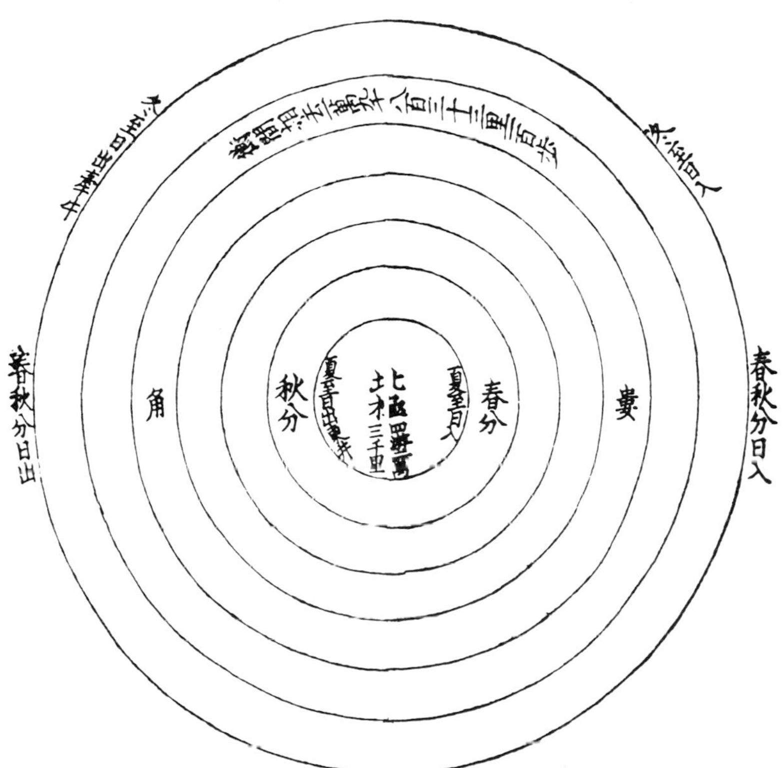

Figure 30. Corrupt diagram of seven *heng*.

Between #D1 and #D2 present texts have a diagram and a short explanatory essay by Zhao Shuang: see p. 46 in Qian's edition. As we have seen, there is some doubt whether the diagram is by Zhao himself, or whether in some form it was already in the text he saw: see above, pages 70 and 184. Whatever the case may be, the traditional diagrams are corrupt: see figure 30. Zhao's essay reads as follows:

> The line drawn in green is the boundary where heaven and earth [apparently] join, the furthest extent of human vision. Heaven is the highest thing of all, and earth is the lowest thing of all. It is not that they really join, rather that the human eye is at

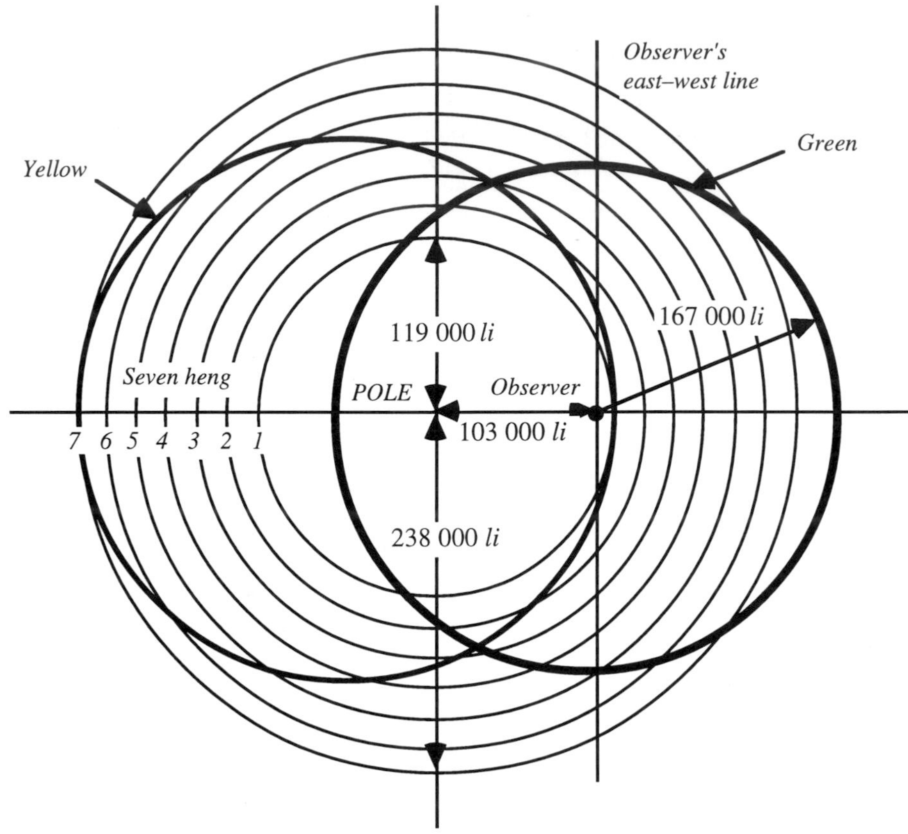

Figure 31. Zhao Shuang's *heng* diagram restored.

the furthest extent of its gaze, and so heaven and earth join.

When the sun comes within the green line we say the sun is coming out. When the sun goes out beyond the green line we say the sun is going in. [In fact] both the inside and outside of the green line are all heaven.

The north pole is exactly at the centre of heaven. What people call east, west, south and north are not fixed places. Everyone calls the place where the sun rises 'east', where the sun is centred 'south', where the sun goes in 'west' and where thc sun is absent 'north'.

Beneath the north pole, the sun is in sight for six months, and out of sight for six months. For the six months from the spring to the autumn equinox the sun is always in sight, while for the six months from the autumn to the spring equinoxes the sun is always out of sight. When the sun is in sight it is day, and when the sun is out of sight it is night. What is called a year is a day and a night below the pole. The line drawn in yellow is the Yellow Road [i.e. the ecliptic]. The 28 lodges are

> laid out upon it, the sun moon and planets move along it. Suppose that the green line is fixed overhead, then if you pierce through the pole and turn it [i.e. the rest of the diagram], then [the two circle] will intersect [at different places].
>
> Our east–west [line] is not the east–west [line] of heaven and earth.
>
> The innermost first [circle] is the [daily] path of the sun at the summer solstice. The middle fourth [circle] is the [daily] path of the sun at the spring and autumn equinoxes. The outer seventh [circle] is the [daily] path of the sun at the winter solstice. Throughout [the sun] follows the Yellow Road. At the winter solstice the sun is in Ox; at the spring equinox it is in Harvester; at the autumn equinox it is in Horn. At the winter solstice it turns from south to north; at the summer solstice it turns from north to south; when it reaches the end it begins once again.

All this is quite clear, and on the basis of the text Zhao's diagram may easily be restored as in figure 31. The system of the *heng* is as described in section #D. We have already heard of the limited range of human vision in paragraph #B25, and the six-month polar day and night are discussed in #B22 to #B24, as well as being implied in #F9. While the Yellow Road is not mentioned explicitly in the main text, its recognition as the basic path of the moving heavenly bodies against the background of the stars is implied in #B20 and #B21. In any case it was a well-known astronomical concept by the time Zhao wrote, even if the universe of the *Zhou bi* cannot represent it as a great circle.

Zhao's original contribution is his concept of a fixed 'window' underneath which the disk of heaven could be rotated in order to show how the sun and other heavenly bodies enter and leave the observer's field of vision. Although this seems to have remained at the level of imagination, the resemblance to a modern planispheric star-finder is interesting. We have already heard of what sounds like a similar device possibly made by Yang Xiong:(page 59), which he destroyed on giving up his belief in the *gai tian* (page 60). In his memorial of AD 92 Jia Kui reports some of his colleagues as stating that 'The star maps [in use today] are "laid out with compasses" *gui fa* 規法', which sounds rather like Zhao's diagram. The *Sui shu* devotes a special section entitled *Gai tu* 蓋圖 'the chariot-cover diagram' to discoid star maps as opposed to those on the surface of a sphere, and it seems they were an established tradition despite the fact that it was known their quantitative predictions had to be treated with care: see *Sui shu* 19, 520.

Appendix Four

The old shadow table

Paragraph #H2 of the present text of the *Zhou bi* lists noon shadows of an eight–*chi* gnomon for each of the 24 'solar seasons' *qi* 氣. The values given correspond precisely to those underlying the dimensions given earlier (section #D) for the seven *heng* circles on which the sun is positioned at the instants of inception of alternate *qi*, the link being the 'one *cun* for a thousand *li*' shadow principle.

It appears, however, that the shadow list seen today is not the one originally seen by Zhao Shuang in the third century AD, but is a new construction by Zhao himself, with added explanatory material in paragraphs #H1, #H3 and #H4. Thus in a comment on #H3, he remarks 'This is Shuang's new method'; he is evidently writing a commentary on a text he has at the least revised himself (Qian edn, 65n). In his final comment on #H4 (Qian edn, 66c ff.) Zhao explains his rejection of the old shadow table in detail:

> The method by which the old shadows were obtained was not in accordance with principle. [Elsewhere in the *Zhou bi*] it is said that at the equinoxes the Yin and Yang shadows are equal, each being 7 *chi* 5 *cun* 5 *fen*, so that the middle *heng* is 75 500 *li* from Zhou. Here however the spring equinox shadow is 7 *chi* 5 *cun* 723 *fen*, and the autumn equinox shadow is 7 *chi* 4 *cun* 262 *fen*, so that there is a discrepancy of 1 *cun* 461 *fen.*[235] Based on this, [the distances of the sun from Zhou] are not equal.
>
> [According to the old text] from Winter Solstice [*qi* #1] to Little Cold [*qi* #2] [the change] is greater by half a days' shadow; from Summer Solstice [*qi* #13] to Little Heat [*qi* #14] [the change] is less by half a day's shadow; from Grain in Ear [*qi* #12] to Summer Solstice is greater by two days' shadow; from Great Snow [*qi* #24] to Winter Solstice is greater by three day's shadow
>
> Further, half a year is 182 5/8 days, but here [the fraction] is given the value 2/4, so that for one day there is obtained 476/730 *cun* of shadow, which is incorrect. The length of a *qi* is not exactly fifteen days, but has a fraction of 7/32 day. If one simply multiplies the daily rate to give [the amount] for a *qi*, the error will be great indeed. The Book of Change says 'An old well has no birds [drinking?]: timely abandonment', meaning that it is time to abandon the use of thirty days [as

[235] As we shall shortly see, these fen are not the usual 1/100 *chi*, but are units of 1/730 *cun*. The shadow table reconstructed in accordance with Zhao's description actually has a spring equinox value of 7 *chi* 5 *cun* 722 *fen*. Since Zhao has calculated the difference between his figures accurately, it seems he may have simply miscopied the spring value.

> the length of two *qi*]. Thus I have made this new method, based on the [true] value of a *qi*, so that its description is simple and its principle is easy, so that high and low are inter-related, and everything moves through a cycle and returns to its original state without discrepancy.

Fortunately there is just enough detail in Zhao's critique to enable us to reconstruct the original *Zhou bi* shadow table which he had deleted in favour of his own creation. In the first place, a half year of 182 2/4 days implies a year of exactly 365 days. Also, using the *Zhou bi*'s usual values for the solstitial shadows, the daily change of shadow length (assumed constant) becomes:

$$\begin{aligned} S &= (13.5 - 1.6)\textit{chi}/(182.5 \text{ days}) \\ &= (119 \textit{ cun})/(182.5 \text{ days}) \\ &= (476 \textit{ cun})/(730 \text{ days}). \end{aligned}$$

This is the value given by Zhao. If, as he notes, the length of a *qi* is taken as exactly 15 days, then the length of the half-year between solstices becomes

$$12 \times 15 \text{ days} = 180 \text{ days}.$$

Thus if we begin from one solstice and add or subtract as appropriate at the rate of 15 × 476/730 *cun* for each *qi*, the value after twelve *qi* have elapsed will differ from the subsequent solstitial value by 2.5 days' worth of shadow, or 2.5 × 476/730 *cun*. The old text of the *Zhou bi* evidently dealt with this by what seems to have been an arbitrary allocation of less or more shadow-change to the *qi* periods on either side of the moments of solstice.[236] For the half-year winter solstice to summer solstice we have an extra half-day's worth for the first *qi* period and an extra two days' worth for period #12; for the next half-year there is half a day's worth less for period #13 and three days' worth more for period #24. In either case, the missing 2.5 days' worth of shadow is made up, and the cycle is completed. On this basis we may reconstruct the original shadow table of the *Zhou bi* as set out in table 6. Shadows are given in units of *chi, cun* and *cun*/730, separated by points in the table.

[236] It naturally occurs to one to total up the number of days'-worth of shadow for each season. This yields spring 92, summer 89.5, autumn 93, winter 90.5. Clearly this has nothing to do with any discovery of the inequality of the seasons, in which case one would have expected figures closer to 94, 92, 89, 90 (Dicks (1970), 24). The pattern here is quite wrong, and in any case the irregularities are squeezed into six *qi* near the solstices rather than attributed to the seasons as a whole. There is no sign that the *qi* were thought of as different lengths, as opposed to having unequal amounts of shadow-change within them.

Table 6. The old shadow table restored

qi	Shadow	Days'-worth	Resultant change
1	13.5.0	15.5	0.10.78
2	12.4.652	15	0.9.570
3	11.5.82	15	0.9.570
4	10.5.242	15	0.9.570
5	9.5.402	15	0.9.570
6	8.5.562	15	0.9.570
7	7.5.722	15	0.9.570
8	6.6.152	15	0.9.570
9	5.6.312	15	0.9.570
10	4.6.472	15	0.9.570
11	3.6.632	15	0.9.570
12	2.7.62	17	0.11.62
13	1.6.0	14.5	0.9.332
14	2.5.332	15	0.9.570
15	3.5.172	15	0.9.570
16	4.5.12	15	0.9.570
17	5.4.582	15	0.9.570
18	6.4.422	15	0.9.570
19	7.4.262	15	0.9.570
20	8.4.102	15	0.9.570
21	9.3.672	15	0.9.570
22	10.3.512	15	0.9.570
23	11.3.352	15	0.9.570
24	12.3.192	18	0.11.538

As I have already mentioned, Zhao's new shadow table is in accordance with the rest of the *Zhou bi*. The old list certainly was not. Although the solstitial values are those used elsewhere in the book, the two differing equinoctial shadows are inconsistent with the assumptions of both section #B and section #D. The use of a 365 day year is also odd. The presence of such material in the *Zhou bi* strengthens the impression that the book is a collection of material of varied provenance rather than the work of a single unifying mind.

References and bibliography

Modern works cited[237]

Anon. (1980) (ed.) *Song ke suan jing liu zhong* 宋刻算經六種 [Six examples of Song dynasty printed editions of mathematical classics], Wenwu chubanshe, Beijing.

Anon. (1987) *Zhongguo tianwenxue shi* 中国天文学史 [A history of Chinese astronomy], Kexue chubanshe, Beijing.

Beer A., Ho Ping-yu, Lu Gwei-djen, Needham J., Pulleyblank E .G. and Thompson G. (1961) 'An 8th-Century Meridian Line: I-HSING's Chain of Gnomons and the Pre-History of the Metric System' *Vistas in Astronomy* 4, 3–28.

Biot E. (1841) 'Traduction et Examen d'un ancien Ouvrage intitule *Tcheou-Pei*, litteralement "Style ou signal dans un circonference"' *Journal Asiatique* 3rd series 11, 593. Supplementary note in (1842) 13, 198.

Chatley H. (1938) '"The Heavenly Cover", a Study in Ancient Chinese Astronomy' *Observatory* 61, 10ff.

Chemla K. (1991) 'Theoretical Aspects of the Chinese Algorithmic Tradition (First to Third Century)' *Historia Scientiarum* no. 42, 75–98.

Chen Jiujin 陈久金 and Chen Meidong 陈美东 (1989) 'Cong Yuanguang lipu ji Mawangdui boshu tianwen ziliao shitan Zhuan Xu li wenti' 从元光历谱及马王堆帛书天文资料试探颛顼历问题 [An investigation of the Zhuan Xu li on the basis of the Yuanguang almanac and the Mawangdui silk manuscript], in *Zhongguo gudai tianwen wenwu lunji* 中国古代天文文物论集 [Collected articles on ancient Chinese astronomical relics], Wenwu chubanshe, Beijing, 83–103.

Chen Meidong 陈美东 (1989) 'Shilun Xi Han louhu de ruogan wenti' 试论西汉漏壶的若干问题 [On some questions relating to Han clepsydras], in *Zhongguo gudai tianwen wenwu lunji* 中国古代天文文物论集 [Collected articles on ancient Chinese astronomical relics], Wenwu chubanshe, Beijing, 137–44.

237 Full and simplified forms of Chinese characters are used in accordance with the original publication.

Chen Zungui 陈遵妫 (1980) *Zhongguo tianwenxue shi* 中国天文学史 [A history of Chinese astronomy], 2 vols., Renmin chubanshe, Shanghai.

Cullen C. (1976) 'A Chinese Eratosthenes of the Flat Earth' *Bulletin of the School of Oriental and African Studies* XXXIX /1, 106–27.

Cullen C. (1980) 'Joseph Needham on Chinese Astronomy' *Past & Present* no. 87, May, 39–53.

Cullen C. (1981) 'Some further points on the *shih*' *Early China* 6, 31–46.

Cullen C. (1982) 'An Eighth Century Chinese Table of Tangents' *Chinese Science* 5, 1–33.

Cullen C. (1993) 'Motivations for scientific change in ancient China' *Journal for the History of Astronomy* 24, 185–203.

Cullen C. and Farrer A. S .L. (1983) 'On the term *hsuan chi* and the flanged trilobate discs' *Bulletin of the School of Oriental and African Studies* XLVI/1, 52–76.

Dicks D. R. (1970) *Early Greek Astronomy to Aristotle*, Cornell, Ithaca, New York.

Ding Fubao 丁福保 and Zhou Yunqing 周云青 (1956) *Sibu zonglu tianwen bian* 四部總錄天文編 [Universal bibliography, astronomy section], Wenwu chubanshe, Beijing.

Dreyer J. L. E. (1906) *A History of Astronomy from Thales to Kepler*, repr. (1953) Dover, New York.

Dull J. L. (1966) *A Historical Introduction to the Apocryphal (Ch'an-wei) texts of the Han Dynasty,* PhD thesis, University of Washington.

Feng Ligui 冯礼贵 (1986) 'Zhoubi suanjing chengshu niandai kao' 周髀算经成书年代考 [On the date of compilation of the *Zhoubi suanjing*], *Guji zhengli yanjiu xuekan* 古籍整理研究学刊 no. 4, 37–41

Fu Daiwie 傅大為 (1988) 'Lun *Zhou bi* yanjiu chuantong de lishi fazhan yu zhuanzhe' 論周髀研究傳統的歷史發展與轉折 [A study of the historical development and transformation of the *Zhou bi* research tradition] *Qinghua xuebao* 18 (new series) 1, 1–41.

Gillon B. S. (1977) 'Introduction, Translation and Discussion of Chao Chun-ch'ing's [= Zhao Junqing = Zhao Shuang] "Notes to the Diagrams of Short Legs and Long Legs and of Squares and Circles"' *Historia Mathematica* 4, 253–93.

Goodrich L. C. and Fang C. (1976) *Dictionary of Ming Biography*, Columbia University Press, New York.

Graham A. C. (1978) *Later Mohist Logic, Ethics and Science*, SOAS, Hong Kong and London.

Graham A. C. (1986) *Yin–Yang and the Nature of Correlative Thinking*, Institute of East Asian Philosophies Occasional Paper and Monograph Series no. 6, Singapore.

Graham A. C. (1989) *Disputers of the Tao*, Open Court, La Salle, Illinois.

Guo Shuchun 郭書春 (1990) *Jiu zhang suan shu* 九章算術 [Nine chapters on the mathematical art], collated text with notes and an introduction, Liaoning Jiaoyu Chubanshe, Shenyang.

Harper D. (1979) 'The Han Cosmic Board (*Shih*)' *Early China* 5, 1–10.

Hashimoto Keizo 橋本敬造 (1980) *Shūhi sankei* 周髀算經 [annotated translation of the *Zhou bi* into modern Japanese] in Yabuuchi (1980), 289–350.

Henderson J. B. (1991) *Scripture, Canon and Commentary: a Comparison of Confucian and Western Exegesis*, Princeton University Press, Princeton.

Hummel A. W. (1943) (ed.) *Eminent Chinese of the Ch'ing Period*, Library of Congress, Washington.

Kalinowski M. (1983) 'Les Instruments Astro-calendériques des Han et la méthode *Liuren*' *Bulletin de l'Ecole Française d'Extrême-Orient* LXXII, 309–419.

Kalinowski M. (1990) 'Le calcul du rayon celeste dans la cosmographie chinoise' *Revue d'Histoire des Sciences* XLIII/1, 5–34.

Karlgren B. (1950) *The Book of Documents*, Museum of Far Eastern Antiquities, Stockholm.

Keegan, D. J. (1988) *The* "Huang-ti Nei-ching": *The structure of the compilation; the significance of the structure*, PhD thesis, University of California, Berkeley.

Kuhn T. S. (1957, repr. 1979) *The Copernican revolution*, Harvard University Press, Cambridge, Mass.

Kuhn T. S. (1970) *The Structure of Scientific Revolutions* (second edn, revised and enlarged), University of Chicago, Chicago.

Lam Lay-Yong and Shen Kangsheng (1984) 'Right-Angled Triangles in Ancient China' *Archive for History of Exact Sciences* 30/2, 87–112.

Li Ling 李零 (1991) '"Shi tu" yu Zhongguo gudaide yuzhou moshi' '式图'与中国古代的宇宙模式 ['Shi-tu' and cosmic models in Ancient China] part 1, *Jiu Zhou Xuekan* 九州学刊 4, no. 1, 5–52.

Li Yan 李儼 and Du Shiran 杜石然 (1976) *Zhongguo gudai shuxue jian shi* 中國古代數學簡史 [Chinese Mathematics: A Concise History], Commercial Press, Hong Kong.

Li Yan and Du Shiran (1987) *Chinese Mathematics: A Concise History,* [English tr. by Crossley J. N. and Lun A. W.-C. of original Chinese text of 1976], Clarendon Press, Oxford.

Lloyd G. E. R. (1973) *Greek Science after Aristotle,* Cambridge.

Lloyd G. E. R. (1987) *The Revolutions of Wisdom: Studies in the Claims and Practice of Ancient Greek Science*, University of California, Berkeley.

Lloyd G. E. R. (1990) *Demystifying mentalities*, Cambridge.

Loewe M. A. N. (1979) *Ways to Paradise: The Chinese Quest for Immortality,* Allen and Unwin, London.

Maeyama Yasukatsu (1975–1976) 'On the Astronomical Data of Ancient China (ca. –100 +200): A Numerical Analysis' *Archives internationale d'histoire des sciences* 25: 247–76, 26: 27–58.

Martzloff J.-C. (1988) *Histoire des Mathematiques Chinoises*, Masson, Paris.

Maspero H. (1929) 'L'Astronomie Chinoise avant les Han' *T'oung Pao* 26, 267.

Mikami Y. (1913) *The Development of Mathematics in China and Japan*, Teubner, Leipzig.

Nakayama S. (1969) *A History of Japanese Astronomy*, Harvard University Press, Cambridge, Mass.

Needham J. (1959) *Science and Civilisation in China*, vol. 3, Cambridge.

Needham J. with Robinson K. G. R. (1962) *Science and Civilisation in China*, vol. IV, part 1, Cambridge.

Neugebauer O. (1975) *A History of Ancient Mathematical Astronomy,* Springer, Berlin.

Newton R. R. (1977) *The Crime of Claudius Ptolemy*, Johns Hopkins University Press, Baltimore and London.

Nōda Churyō 能田忠亮 (1933) *Shūhi sankei no kenkyū* 周髀算經の研究[An enquiry concerning the *Zhou bi suan jing*], Tōhō Bunka Gakuin, Kyōto.

Qian Baocong 錢寶琮(1924) '*Zhou bi suan jing* kao' 周髀算經考 [A study of the *Zhou bi suan jing*], *Kexue* 科學 14, part 1, repr. in Qian (1983), 119–36.

Qian Baocong 钱宝琮 (1958) 'Gai tian shuo yuanliu kao' 盖天说源流考 [On the origins of the *gai tian* theory], *Kexue shi jikan* 科学史集刊 1, March, repr. in Qian (1983), 377–403.

Qian Baocong 錢寶琮 (1963) (ed.) *Suanjing shishu* 算經十書 [The Ten Mathematical Classics], Kexue chubanshe, Beijing.

Qian Baocong 钱宝琮 (1964) *Zhongguo shuxue shi* 中国数学史 [A history of Chinese mathematics], Kexue chubanshe, Beijing.

Qian Baocong 钱宝琮 (1983) *Qian Baocong kexue shi lunwen xuanji* 钱宝琮科学史论文选集 [Selected works on the history of science by Qian Baocong], Kexue chubanshe, Beijing.

Rodzinski W. (1984) *The Walled Kingdom: A History of China from 2000 BC to the Present,* Fontana, London.

Saussure F. de (1983) *Course in General Linguistics*, [English tr. Roy Harris], Duckworth, London.

Schafer E. H. (1977) *Pacing the Void: T'ang Approaches to the Stars*, University of California, Berkeley.

Sivin N. (1969) *Cosmos and computation in Early Chinese Mathematical Astronomy*, Brill, Leiden.

Sivin N. (1987) *Traditional Medicine in Contemporary China*, University of Michigan, Ann Arbor.

Sivin N. (1992) 'Ruminations on the Tao and its Disputers' *Philosophy East and West* vol. 42, no. 1, 21–9.

Smart W. M. (1979) *Spherical Astronomy* (6th edn) Cambridge.

Swetz F. J. and Kao T. I. (1977) *Was Pythagoras Chinese? An Examination of Right Triangle Theory in Ancient China*, Pennsylvania State University Press, University Park and London.

Toomer G. J. (1984) (translator) *Ptolemy's Almagest*, Duckworth, London

Waerden B. L. van der (1983) *Geometry and Algebra in Ancient Civilisations*, Springer, Berlin.

Wagner D. W. (1985) 'A Proof of the Pythagorean theorem by Liu Hui (Third Century AD)' *Historia Mathematica* 12, 71–3.

Wang Jianmin 王健民 and Liu Jinyi 刘金沂 (1989) 'Xi Han Ruyin Hou mu chutu yuanpan shang ershiba xiu gu judu de yanjiu' 西汉汝阴侯墓出土圆盘上二十八宿古距度的研究 [On the old extensions of the 28 lodges as seen on the disc excavated from the Western Han tomb of the Marquis of Ruyin], in *Zhongguo gudai tianwen wenwu lunji* 中国古代天文文物论集 [Collected articles on ancient Chinese astronomical relics], Wenwu chubanshe, Beijing, 59–68.

Wylie A. (1897) *Notes on Chinese Literature*, Shanghai, reprinted (1964) Literature House, Taipei.

Yabuuchi Kiyoshi 藪內清 (1969) *Chūgoku no Temmon Rekihō* 中国の天文曆法 [Chinese mathematical astronomy], Heimarusha, Tokyo.

Yabuuchi Kiyoshi 藪內清 (1980) (ed.) *Kagaku no meicho* 科学の名著 [Great books of science] vol. 2, *Chūgoku Temmongaku, sūgaku shū* 中国天文学數学集 [Chinese astronomy and mathematics], Asahi shuppansha, Tokyo.

Selected pre-modern texts cited

Beitang shuchao 北堂書鈔 encyclopaedia by Yu Shinan 虞世南, *c*. AD 639, Taipei 1962 repr.

Chuxue ji 初學記 encyclopaedia by Xu Jian 徐堅, *c*. AD 700, Beijing 1962 edn.

Dai Zhen wenji 戴震文集[Collected works of Dai Zhen, 1724–1777], Beijing 1963 edn.

Fa yan 法言 'Exemplary Sayings' by Yang Xiong 揚雄, *c*. AD 9, *Sibu congkan* edn.

Han shu 漢書 History of Western Han dynasty by Ban Gu 班固, largely complete on his death in AD 92, Beijing 1962 edn.

Hou Han shu 後漢書 History of Eastern Han dynasty by Fan Ye 范曄, *c*. AD 450, Beijing 1963 edn.

Huai nan zi 淮南子 Compendium of learning assembled under the patronage of Liu An 劉安, Prince of Huai Nan, completed by 139 BC, *Sibu congkan* edn.

Huang di nei jing 黃帝內經 'Inner canon of the Yellow Emperor' a composite work assembled but not definitively fixed in content around AD 100, Shanghai 1955 edn.

Jin shu 晉書 History of the Jin dynasty (AD 265–419) by Fang Xuanling 房玄齡, *c.* AD 648, Beijing 1974 edn.

Jiu Tang shu 舊唐書 Old History of the Tang dynasty (AD 618–906) by Liu Xu 劉煦 AD 945, Beijing 1975 edn.

Jiu zhang suan shu 九章算術 'Nine Chapters on the Mathematical Art' compiled *c.* AD 100, edn of Guo Shuchun 郭書春, Shenyang 1990.

Kao gong ji 考工記 'Record of the Artificers' a document of the late Warring States period, *c.* 300 BC, now incorporated in the *Zhou li.*

Lü shi chun qiu 呂氏春秋 'Mr. Lü's Spring and Autumn Annals' compendium of learning completed *c*. 239 BC, *Sibu congkan* edn.

Lun heng 論衡 'Discourses weighed in the balance' by Wang Chong 王充, *c*. AD 75, *Hanwei congshu* edn.

Qinding siku quanshu 欽定四庫全書 'Imperially commissioned complete [manuscript] collection of the four categories of literature', completed 1782, photographically printed Shanghai 1983.

San guo zhi 三國志 History of the Three Kingdoms period (AD 221–280) by Chen Shou 陳壽, *c*. AD 290, Beijing 1962 edn.

Shi ji 史記 'Records of the Historian' by Sima Qian 司馬遷, *c.* 90 BC, Beijing 1962 edn.

Shu jing 書經 'Book of Documents' complex and heterogeneous collection of material, some possibly as early as early first millennium BC, with commentaries and subcommentaries in *Shisan jing zhushu* 十三經注疏, repr. Taipei 1972 from edition of 1815.

Shuo yuan 説苑 Collection of moral tales and political admonitions by Liu Xiang 劉向, *c.* 17 BC, *Sibu congkan* edn.

Siku quanshu: see *Qinding siku quanshu*

Song shu 宋書 History of the Liu Song dynasty (AD 420–479) by Shen Yue 沈約, AD 488 (monographs added after AD 500), Beijing 1974 edn.

Sui shu 隋書 History of the Sui dynasty (AD 581–617) by Wei Cheng 魏徵, AD 636 (monographs added AD 656), Beijing 1973 edn.

Taiping yulan 太平御覽 Encyclopaedia by Li Fan 李昉 and others, AD 983. Shanghai 1960 repr. of *Sibu congkan* edn.

Wei shu 魏書 History of Toba Wei dynasty (AD 386–550) by Wei Shou 魏收, AD 554 revised 572, Beijing 1974 edn.

Xia xiao zheng 夏小正 'Lesser Annuary of the Xia [dynasty]' pre-Qin, *Congshu jicheng* edn.

Xin Tang shu 新唐書 New History of the Tang dynasty (AD 618–906) by Ouyang Xiu 歐陽修, *c*. AD 1060, Beijing 1975 edn.

Yi jing 易經 'Book of Change' originally a manual of divination (pre-Qin), with later accretions dating as late as W. Han, *Sibu congkan* edn.

Yue ling 月令 'Monthly Ordinances' of uncertain date, now incorporated in *Lü shi chun qiu.*

Zhou li 周禮 'Ritual of the Zhou dynasty' late Warring States, with commentaries and subcommentaries in *Shisan jing zhushu* 十三經注疏, repr. Taipei 1972 from edition of 1815.

Zuo zhuan 左傳 Historical narratives from Spring and Autumn Period, with commentaries and subcommentaries in *Shisan jing zhushu* 十三經注疏, repr. Taipei 1972 from edition of 1815.

Index